Shades of Darkness

Published by
American Book Business Press™
An imprint of American Book Publishing
P.O. Box 65624
Salt Lake City, UT 84165
www.american-book.com

Printed in the United States of America on acid-free paper.

Shades of Darkness: California's Energy **Crisis**
Designed by Jana Rade, design@american-book.com

Publisher's Note: *American Book Publishing relies on the author's integrity of research and attribution; each statement has not been investigated to determine if it has been accurately made. The author and publisher specifically disclaim any responsibility for any liability, loss, or risk, personal or otherwise, which is incurred as a consequence, directly or indirectly, of the use and application of any of the contents of this book. In such situations where medical, legal, or other professional services may apply, please seek the advice of such professionals directly.*

ISBN-13: 978-1-58982-425-6
ISBN-10: 1-58982-425-3

Library of Congress Cataloging-in-Publication Data
Dulude, Bill, 1927-
Shades of darkness : California's energy crisis / Bill Dulude.
p. cm.
ISBN-13: 978-1-58982-425-6
ISBN-10: 1-58982-425-3
1. Electric utilities--California. 2. Electric power consumption--California. 3. Energy policy--California. 4. Energy industries--California. I. Title.
HD9685.U6C327 2008
333.793'209794--dc22

2008030529

Special Sales

These books are available at special discounts for bulk purchases. Special editions, including personalized covers, excerpts of existing books, and corporate imprints, can be created in large quantities for special needs. For more information e-mail info@american-book.com.

Shades of Darkness: California's Energy Crisis

Bill Dulude

Dedication

I dedicate this book to my wife Bobbie, who has been patient and completely supportive of me in my endeavors in writing this book and, for the past fifty-two years, has been my friend, my companion, my counselor, and my lover.

"Government should only do those things
the people cannot do for themselves."

Ronald W. Reagan, June 23, 1971

Amen!

Foreword

Shades of Darkness: California's Energy Crisis is a fascinating but an acutely depressing read. It is a document that deserves to be in every university library, not only in America but Europe too. The California crisis is echoed across the globe. California's crisis is similar to the crises facing generating undertakings throughout the industrialized world, and the author has graphically highlighted the California failure. Every U.S. state and every industrialized nation can learn from California's mistakes.

The ill-managed Enron did possibly get one thing right: its prediction that the world's growing energy needs will require the construction of more electrical generating capacity than has been built in the past one hundred years.

California's generation and pricing problems are far from unique. Britain has similar problems where privatization, after years of nationalization, has proved to be a huge disappointment and where the aspiration of the shareholders has taken precedence over the needs of the nation as a whole. Since privatization in 1988, there has been little or no investment in new plants. Reliance has been placed on generation from cheap supplies of North Sea gas. The widespread belief that new resources would be found in the Irish Sea and off the coast of Scotland has proved to be unfounded. Without new discoveries, the North Sea gas supplies, as it was predicted back in 1980, will run out by 2012—if not sooner. For once the predictions were spot on.

There is now a pipeline from Norway, 700 miles away, bringing gas to our generation stations, a pipeline which was hurriedly brought into use in September 2006, several months early. In addition, contracts to ship in liquid gas from sources such as Algeria have already been signed with dubious commercial terms—contracts that have to be honored.

Belatedly, it is now realized that England has gone from energy over producer to an energy importer, importing from countries such as Norway and Africa and the Gulf States. Some energy is brought in from France, where nuclear-generated power comes from aging power stations that still reign supreme.

Worldwide forecasts have long been ignored, along with a failure to grasp a whole series of fundamentals. As the boss of Pepsi Cola said of its rival after the expensive disaster of the launch of New Coke, "If you don't understand the product, you ain't goin' nowhere." Electricity is the product we are concerned with, and sadly our basic understanding of it is not a great deal better than when Faraday famously announced to his students that they should not waste time studying electricity, as the applications for its use would always be very limited.

What would Faraday have thought of the nature-related accident that brought a city of several million to a startling halt in 2002 (as I recall)? An ice storm destroyed the transmission towers in Montreal, and the city went without electricity for almost a week. Amazingly the death toll was small, but the economy collapsed, and all forms of human life descended to bare subsistence levels. Predictions of it happening had been ignored, and the cost of undergrounding the cables was considered to be too expensive—the cost would have been one-fifth of what it eventually cost the city. Nothing has been learned. The transmission towers are all there again, and in the next decade the weather experts confidently predict that with climate change it will happen again—and not just in one city in Canada.

It was recognized half a century ago that fossil fuels were finite that there was a limit to the building of hydro dams in the Western Hemisphere, and that wind and tidal power were not the most cost-effective form of generation. The best way of generating electricity was thought to be nuclear power. However, in the haste to build the generating stations, the engineering problems were never properly examined, and few bothered to study the difficulties associated with decommissioning until it was far too late.

The author's list of the various methods of generating renewable energy presents an opportunity, but even if all were to be adopted, it would only account for a fraction of the growing needs. There is an irony, too, in that the exponents of renewable resources are for the most part the same people who make every effort to prevent the building of hydro dams because of the perceived adverse effect on nature. In Canada, a hydro project was scrapped because it threatened the only known site of a species of a spider that you could barely see without a magnifying glass. However, it has been revived as someone has come up with the brilliant idea of moving the spiders to a new friendly habitat—if they can find them.

In China, they are completing the world's largest hydro scheme only to discover that it comes at a huge cost in terms of human displacement and long-term environmental damage. And it will not entirely solve the problem of river flooding that has plagued the nation.

I was pleased to read that the author is a supporter of geothermal generation. Geothermal energy is an opportunity that has been largely ignored. A project was proposed for Britain, but it was in an area of "outstanding natural beauty," and the environmental lobby protested. There was to have been a public enquiry, but that was overshadowed by the enquiry concerning the nuclear station at Sizewell, an enquiry that lasted for three years and cost almost as much as building the station. The thought of another enquiry was too much for successive governments until, and with the impetus of panic, the proposal of two new nuclear stations was floated this year.

Here in Britain, a plan to build some two thousand wind turbines has triggered vociferous protests from those living as far as a mile away claiming that their comfort would be destroyed because of the twenty-four-hour noise made by the turning vanes. Another environmental group complained because the vanes were unsightly and proposed that they should all be built out at sea—on redundant North Sea oil platforms! Solar power has been criticized for a variety of reasons, not least its failure to produce the anticipated kwh returns and a higher cost of maintenance than was ever envisaged. Methane electricity generators have raised serious safety issues following an explosion that was at first credited to a terrorist bomb.

In the panic to find and fund energy sources, oil shale was seen as one solution. Venezuela and Canada were considered to be the ideal sources. To give it street credibility, the name Orimulsion was coined, but it proved to be a substance so toxic that it has to be shipped in special vessels. The sulfur content of Orimulsion is so high that the installations of very expensive filters are required. In a trial at Richborough Power Station here in England, a station which formerly burnt crude oil, one of the filters failed, and the company faced a $6 million lawsuit as the fallout stripped the paint from imported new cars awaiting distribution to dealers, and human sufferers filled the local hospitals with lung complaints. The government's Health and Safety Executive scrapped the scheme and cancelled an order for two specially designed tankers from Japan, resulting in another lawsuit.

The energy industry worldwide had been praying for a miracle. At a European Power Convention in Geneva in the late '60s, one U.S. scientist stated confidently that by the year 2000, most homes would have their own generators and mini-power stations, and failure-prone countrywide transmission lines would be redundant. His claim was that the domestic consumer would buy a power generator in the same way as buying a household heating boiler or a washing machine or a television. The fuel source would be hydrogen. His picture is not quite as fantastical or risible as it might seem; since the year 2006, we have seen the launch in the U.S. of a family-sized automobile running on electricity created by hydrogen fuel cells.

This Alice in Wonderland scenario of "Jam Tomorrow" has colored the thinking of engineers and scientists who long ago decided to patch up and extend the life of old generating methods until the coming of entirely new scientific and supposedly safe nuclear technology. When that did not arrive quite as expected, then they put their betting money on the coming of nuclear fusion or the results of research into particle physics.

The proponents of particle physics will tell you that their research has absolutely nothing to do with electricity generation, but in order to get more funding, they have to pretend it has. They then look shifty and whisper that once they have cracked the mysteries of the universe and harnessed the useful electron, there should be enough free energy for everyone. The problem was going to be how to harness it, but that was one for the engineers.

The United States Congress in the 1980s spent $2 billion on the Superconducting Supercollider at Waxahachie, Texas. The plan was to fire atomic particles through a tunnel 50 miles long and produce a staggering 100 trillion volts of energy. Once having cracked that problem, it was thought feasible to turn that energy into electrical power using the engineering technology learned from the nuclear industry.

Then came the doubts. Particle accelerators built in South Dakota and Illinois suffered the same fate as Waxahachie after critics predicted that it would be possible to create 'strange matter' more stable than ordinary matter and even, heaven help us, black holes capable of destroying the earth. Fortunately for science, the Fermilab near Chicago continues to operate and privately sees electricity generation as a practical possibility, although at a very high price.

In Europe the effort continues with the Large Hadron Collider. Over two thousand scientists from thirty-four countries are working on this project, scheduled to begin operating in November 2007. So far it has cost $8 billion to build, and if it works—and the only thing that is certain in atomic physics is that nothing is certain—then 14 trillion volts of energy will be produced. The French are the biggest supporters of this effort since they have to find some means in the next ten years to replace their pressurized water reactors, most of which have long passed their sell-by date. Already they have a program of new updated versions of the present pressurized water reactors.

I enjoyed reading about the U.S. contracts and pricing structures. It has always been my belief that you cannot wholesale and sign contracts to supply a product that cannot be stored. The economics of supplying a product on instant demand with unpredictable peaks and troughs poses an almost unsolvable problem, with approximately 20 percent of the electricity generated being wasted. Pumped-storage hydro stations, such as the one we have here in Wales, can meet sudden demand, but they are not cost effective since electricity is required to pump the water back up into the upper dam. In an effort to balance the load on the system, off-peak tariffs were tried in the UK. They, too, proved not to be cost effective and have largely been abandoned.

Competition among the various undertakings is far from achieving the predicted price reductions and instead has had the opposite effect, ignoring the politicians' assurances that the price of electricity would never increase above the cost of inflation. In order to keep the shareholders happy, there has been little investment in new plants. Fluctuations in the wholesale price have led inevitably to arbitrary price rises. British Gas has increased energy prices by 85 percent in less than two years. Npower, with over 6 million customers, announced in the summer of 2006 that retail gas bills would go up by an average of 17 percent and electricity by 10 percent, this in the same month that wholesale gas prices dropped by 18 percent and electricity by 20 percent. No immediate reductions are envisaged, and the view of one economist, shared by many, is, "The companies want to boost profits while they can."

Another cause of consumer fury is that the government, in its quest to produce revenue, has added a sales tax to every energy bill.

Consumer protection is another sore point. PowerGen, the company that owns the generating stations, has been fined over $500,000 by the industry's regulator OFGEM for unfair competition and price fixing. However, the German Company E.ON that owns PowerGen says it has no intention of paying and will continue its illegal pricing policies unless there is a judgment by the European Court—which could take years.

This highlights another difficulty. Firms in Europe, Asia, and America, despite earlier promises, now own the majority of the UK power companies. The one that remains stoutly in British ownership is the one owning the nuclear power stations; this is only because nobody wants to buy that loss-making company.

It comes down to this: Is anyone going to learn or act on the author's brilliant exposition of the failures, the chicanery, and the on-going problems? I had a very enjoyable read and found myself in agreement with most of the author's history and the conclusions to be drawn, but I just don't see the author being called before a senate subcommittee as an expert witness. His testimony would probably be too embarrassing. I hope I am wrong.

What the author has written deserves a wide audience. But I have become very cynical over the years, as I have lived and worked through power crises here in Britain and come to regard both politicians and energy "experts" with grave suspicion. I tend to sympathize with economists who urge for the price per kwh to be raised by 30 percent in stages over the next five years to ensure investment in new plants and improved transmission systems, but with the U.S. economy and ours so dependent on electric power, I can't see it happening. Even so, that revenue would barely be enough to fund the investment in environmentally-friendly generation.

The book is quite a thesis. The author deserves a PhD for his study in incompetence and inefficiency and chicanery. No question that California has an energy crisis with no short-term solution. Maybe it can take comfort in the fact that its problems are common to every developed nation today.

Conservation and the development of more energy efficient appliances, while laudable objectives, are not the solution, and any contribution will be small. Heat pumps instead of air-conditioning units have still not lived up to expectations in cost saving and performance. Solar tiles have their drawbacks since their efficiency declines over time with dirt and pollution, and few can be bothered to clean their roofs regularly.

Even a powerful committee such as the author has envisaged will not provide the answer in the long term unless they are given draconian legislative powers—which are the anathema to the American political system. Stable pricing structures are going to be difficult to apply in the present economic and political climate. Nothing will happen until the lights start going out and the huge American freezers start defrosting. The U.S. beer drinkers don't like warm beer, and that will—as Dr. Johnson said about hanging—concentrate their minds wonderfully. Then there will be action.

The author's impressive study deserves to be widely distributed and read and debated.

Anthony Byers
Year 2006

****

Mr. Byers retired as Chief Press Officer of the Electricity Council, which was the controlling body of England's nationalized electricity industry up until its privatization. He has authored a number of published articles and books on electrical subjects.

Acknowledgments

An immeasurable amount of the information, including quotations, for *Shades of Darkness* were obtained from articles in my subscriptions to the *Sacramento Bee*, Placerville's *Mountain Democrat*, and the *U.S. News & World Report* magazine in the period from September 2000 to completion of the book. Additional information was sought from the players in the crisis. Some were very cooperative, others were not cooperative, and some were completely silent. Those who were cooperative and informative included:

- ✓ Gregg Fishman, spokesman for the ISO, was always prompt and extremely informative and helped my understanding of the ISO's operations.
- ✓ Malcolm McCay, Sempra Energy (SDG&E), was another player with prompt and informative responses to my inquiries.
- ✓ Richard Wheatley, Reliant Resources, gave a good critique of the introduction to Reliant along with being openly informative.
- ✓ Patrick Mullen, Duke Energy. Apparently my inquiry floated around Duke Energy for a while and finally wound up in the hands of Patrick, who promptly called me and provided me with a lot of good information.
- ✓ Susanne Garfield-Jones, CEC, was always prompt with responses to my inquires.
- ✓ Sheri Inouye, PUC, was prompt, and if she didn't have the answers, she would get them for me.

- ✓ Nicholas A. Flegel, a law student and son of a very good friend.
- ✓ George Bill, a very good friend. George made an outstanding contribution by editing the original manuscript of my book.
- ✓ Nicholas Fox, editor at American Book Publishing, was extremely helpful in organizing my book. He provided considerable general guidance. Unfortunately he was unable to dedicate adequate time to the book and subsequently took leave of editorial participation.
- ✓ Pamela Peterson eventually became the editor of my book. Pam did a yeoman's job of editing my book. She was extremely diligent in her pursuit of her work. She kept me honest on many things, and I enjoyed working with her.

Without the help from these folks, I may not have been able to write and publish the book. I owe them all a debt of gratitude, and they do have my sincere appreciation.

Mirant was one of the players I sent an inquiry to. I received a telephone call from them telling me they could not help me, and it was my sense that the caller could not get off the telephone fast enough. I did appreciate the response—at least they stepped up to the plate and said they could not help.

Then there were those that would not respond at all, and they included ARB, FERC, PG&E, SCE, Christopher Weare of PPIC, Frank Wolak, a Stanford professor, Calpine, Dynergy, Enron, and Williams/AES. It is unfortunate they failed to respond. I believe they could have made worthy contributions to the book, and I believe it would have been in their best interest to respond.

Table of Contents

Abbreviations

AECA	Agriculture Energy Consumers Association
AEP	American Electric Power Company
AFC	Application for Certification [California]
AGA	American Gas Association
AMP	Automatic Mitigation Procedures
ANWR	Asiatic National Wildlife Reserve
ARB	Air Resources Board
AWEA	American Wind Energy Association
BP	British Petroleum
BPA	Bonneville Power Association
BLM	Bureau of Land Management
CACE	California Alliance for Competitive Energy
CBEA	California Biomass Energy Alliance
CEC	California Energy Commission
CEOB	California Electricity Oversight Board
CERA	Cambridge Energy Research Associates
CERS	California Energy Resources Scheduling
CFA	California Forestry Association
CFB	Circulating Fluidized Bed
CFTC	Commodity Futures Trading Commission
CLECA	California Large Energy Consumers Association
CMTA	California Manufacturing & Technology Association
CPCFA (or CPA)	California Power & Conservation Financing Authority
CPIRG	California Public Interest Research Group
CRCDC	California Energy Resources Conservation and

	Development Commission
CRE	Californians for Renewable Energy
CTC	Competitive Transition Charge
CURE	California Unions for Reliable Energy
DMA	Department of Market Analysis [ISO]
DWR	Department of Water Resources
EIA	Energy Information Administration
ESP	Electric Service Providers
FACTS	Flexible AC Transmission System
FERC	Federal Energy Regulatory Commission
FPA	Federal Power Act
FPC	Federal Power Commission
FTA	Fixed Transition Amount
FTCR	Foundation for Taxpayers and Consumers Rights
GAO	General Accounting Office [Congress's investigative arm]
GMC	Grid Management Charge
HVDC	High Voltage Direct Current
IEP	Independent Energy Producers
IIR	Industrial Information Resources
IOU	Investor Owned Utilities
IPPA	Independent Power Producers Association
ISO	Independent Service Operator
kw	kilowatt
kwh	kilowatt hour
LADWP	Los Angeles Department of Water & Power
LAFCO	Local Agency Formation Commission
LFG	Landfill Gas
LMP	Locational Marginal Pricing [ISO]
LNG	Liquefied Natural Gas
MBTU	Million British Thermal Units
MD02	Market Design 2002 [ISO Comprehensive Market Design Proposal]
MSC	Market Surveillance Committee [ISO]
munies	municipal utilities
Mw	megawatt
mwh	megawatt hour
NARUC	National Association of Regulatory Utility

	Commissioners
NERC	North American Electric Reliability Council
NIMBY	Not in My Back Yard
NOI	Notice of Intention [California]
NRC	Nuclear Regulatory Commission
NRRI	National Regulatory Research Institute
O&I	Oversight and Investigations Program [ISO]
OCC	Official Committee of Unsecured Creditors
OPEC	Oil Producing Exporting Companies
PBMR	Pebble Bed Modular Reactor
PPA	Public Power Association
PPIC	Public Policy Institute of California
PUC	[California] Public Utilities Commission
PUCHA	Public Utility Company Holding Act
PURPA	Public Utility Regulatory Policies Act
PX	Power Exchange [California]
QF	Qualifying Facility
RAR	Resource Adequacy Requirement [ISO]
RCRC	Regional Council of Rural Counties
RPS	Renewable Portfolio Standard
RRBMA	Rate Reduction Bond Memorandum Account
RTO	Regional Transmission Organizations
SCR	Selective Catalytic Reduction
SEC	Security Exchange Commission
SMUD	Sacramento Municipal Utility District
SOCAL	Southern California Gas Company
TOU	time of use
TURN	The Utility Resource Network
UC	University of California-Berkeley
UCAN	Utility Consumer Action Network
UCC	Unsecured Creditors Committee
UCEI	University of California Energy Institute
UCLA	University of California at Los Angeles
WAPA	Western Area Power Administration
WPESC	Western Power Exchange Steering Committee
WSCC	Western System Coordinating Council
WPTF	Western Power Trading Forum

I. Sundown—California's Energy Crisis

There is one essential thread running through all of our lives, a thread that is the essence of our vibrant economy—a thread that we, the consumers, take for granted and have become totally and complacently dependent upon—that thread is electrical energy. We take electrical energy for granted when:

We arrive home at night and flip a switch on the wall to have light.

We pick up the telephone, dial, and expect a response from the individual on the other end of the line.

We open the refrigerator seeking something for dinner and cook it on our electric stove or in the microwave.

We load the laundry into the washer and dryer, push a button, and walk away.

We go to the computer to check for or to send email.

Our doctor orders an X-ray or MRI or a scan for us, and we are oblivious to what electrical energy has done to bring us all the miracle benefits of modern medicine.

The list is endless.

Over the past twenty-five years, California's electrical systems have not received the attention they have deserved. California has been comfortably dependent on the availability of electrical energy from its neighboring states to meet its insatiable appetite for electrical energy, apparently not realizing its neighboring states' energy was never a dedicated source of energy for California. It was, in reality, surplus energy those states had available at the time. No consideration was ever given to the possibility that one day those neighbors just might have a need for their own surplus energy to meet their increasing energy demands. That day came in 2000 and early 2001; it was precipitated by a drought in the Pacific Northwest. California then was in a precarious position because California did not have the generation capacity necessary to be self-reliant.

Consequently, California experienced several blackouts in the year 2001. A considerable amount of California's energy had come from our northwestern coastal states—Washington's and Oregon's hydroelectric facilities—and those states experienced severe droughts in 2000 and 2001. The shortage was intensified by limited natural gas pipeline capacity across the Arizona-California border and by restricted electrical transmission thermal capacity along Path 15, an 85-mile stretch of electrical transmission lines in the Central Valley that connect Southern California with the northern part of the state—a notorious transmission bottleneck.

California's maximum demand in the early 2000s had hovered around 45,000 megawatts (Mw). Through conservation and the start-up of several new power plants, and with the arrival of fall weather, the state was able to ease itself out of the energy crisis. Underneath it, however, some mischievous electrical energy and natural gas suppliers had manipulated the supply markets; reaping excessive profits at the state's expense, and this had to some extent exacerbated the energy shortage.

In 1996 the state's three major electrical utilities—Pacific Gas & Electric, Southern California Edison, and San Diego Gas & Electric—supplied 77 percent of the state's customers' electrical energy requirements (kwh) and provided 75 percent of their electrical energy demand (kw). Two major municipal utilities—Los Angeles Department of Water and Sacramento Municipal Utility District—supplied 15 percent of the state's power requirements. The remaining balance was provided by various smaller municipal utilities and cooperatives. California electric utility service areas are shown in Figure 1, below:

Map of Electric Utility Districts in California - 1996

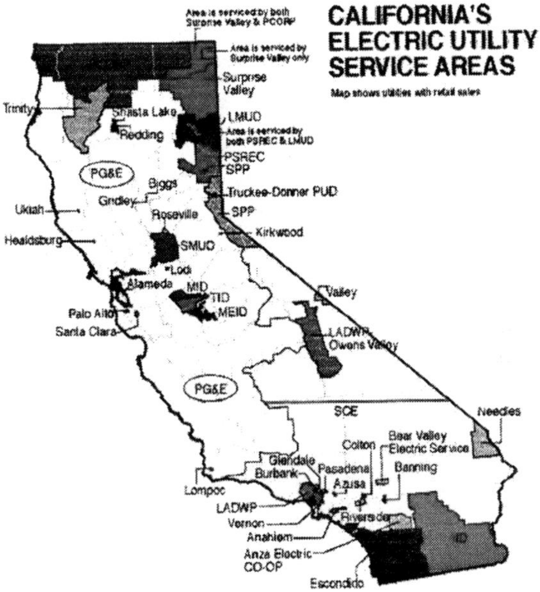

Figure 1.
California Electric Utility Areas
Courtesy of the CEC[1]

---

[1] Abbreviations: CSC: Santa Clara Electric Department; LADWP: Los Angeles Department of Water & Power; MEID: Merced Irrigation District; PCORP: Pacific Power (PacifiCorp); PSEREC: Plumas-Sierra Rural Electric Cooperative; SDG&E: San Diego Gas & Electric; SPP: Sierra-Pacific Power; IID: Imperial Irrigation District; MID: Modesto Irrigation District; PG&E: Pacific Gas & Electric; SCE: Southern California Edison; SMUD: Sacramento Municipal Utility District; TID: Turlock Irrigation District

California did have a low annual rate of consumption relative to the nation's consumption—6,400 kwh per individual versus 11,900 kwh per individual nationally. However, in the early 1990s California's electrical retail rates were, and remain, among the highest in the nation.

The 1970s energy shortage, created by OPEC's manipulation of the oil markets, caused the federal government to pass the Public Utility Act of 1978. This act allowed non-utility entities to build renewable energy generating facilities, which were known as Qualifying Facilities (QFs). The act provided tax benefits but required the utilities (as captive buyers) to purchase QF power. The costs of QF power, along with nuclear power plant costs, were responsible for California retail rates being the highest in the nation.

Prior to the passage of California's utility deregulation act AB 1890, California's three major utilities operated as regulated entities with the responsibility to provide electrical energy to their customers, either though generation or wholesale purchase for resale. That requirement included delivery of electrical energy through high-voltage transmission systems to the areas of demand as well as the distribution of electrical energy within those areas. California's transmission systems are shown in Figure 2, below:

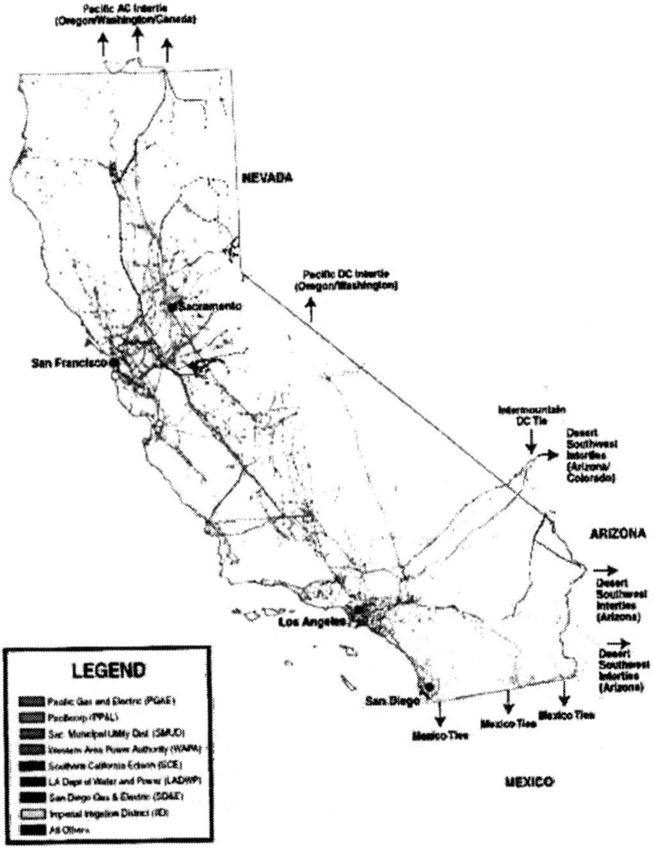

Figure 2.
California Electrical Transmission Systems
Courtesy of the CEC

The retail cost to the customer was, and remains, regulated by the California Public Utility Commission (PUC), which has the dual responsibility of holding down consumer energy prices and assuring the utilities they will recover their costs with a reasonable return on investment in order to maintain viable businesses. The Federal Energy Regulatory Commission (FERC), as a result of deregulation, now regulates the wholesale cost of energy.

Deregulation

In 1992 the federal government, under the supervision of FERC, began promoting deregulation of the electrical industry. The objective was to make the electrical energy market more competitive. In 1996 California complied by passing AB 1890, which was signed into law by then Governor Pete Wilson, becoming effective in 1998. The objective of the measure was to lower energy retail rates to the consumer. The AB 1890 document read in part, "SUMMARY: Restructures the electrical services industry in California in order to transition to competitive markets by December 31, 2001, to lower the cost of electricity, retain and attract jobs, and to reduce power outages."

Obviously, the objectives were not fulfilled. The framers of the bill didn't consider that suppliers of energy would build only enough generating plants to supply energy they could sell at a price that would assure them a satisfactory return on investment. Reserve power was not considered, and a capacity component was not included in the rate structure.

When California's utility deregulation was actually initiated in 1998, the bill required (1) all three private utilities to sell their fossil fuel power generating facilities and (2) removal of PUC authority over wholesale prices of electrical energy sold to private utilities. Additionally, it established the Power Exchange (PX) as the state's central market for trading wholesale energy between the suppliers and the utilities. The utilities no longer had the freedom to negotiate or purchase energy from suppliers. It established an Independent Service Operator (ISO), supervised by FERC, to manage electrical energy and gave it the authority and responsibility to coordinate energy purchases and the dispatching of energy through the transmission systems to the various utilities' receiving points. The ISO service areas are shown in Figure 3, below:

Figure 3.
The California ISO Control Area
Courtesy of the CEC

The utilities retained their authority and management of their distribution systems and their remaining generation facilities. The retail price of the electrical energy remained regulated by the PUC; it was reduced by 10 percent and was frozen with the implementation of AB 1890. The frozen 10 percent reduction applied only to PG&E and SCE; however, it allowed those utilities to recover their "stranded" costs of the transition.[2] SDG&E retail rates were not frozen. However, they were eventually returned to PUC's rate control with the passage of AB 265.

The concept of "direct access" was established, allowing customers to purchase their energy from their choice of suppliers or marketers and not necessarily from the utilities.

The framers of deregulation anticipated that power generators would flock to California, hustling to build new generating facilities—the framers expected it to result in a highly competitive electrical energy market along with low retail energy prices. Five out-of-state generators did enter the California market by purchasing some of the utilities' generating facilities. Calpine Corporation was the only in-state company to participate in such purchases. Several of those generators did plan to construct additional power-generating facilities, and they submitted permit applications to construct additional facilities. It was then they realized they had to conform to the permit regulations of the California Energy Commission (CEC) along with the requirements of the Air Resources Board (ARB). The CEC permit process, initiated in the 1970s, originally required the applicant to file a Notice of Intention (NOI) to build a power plant after the applicant had selected a site and commenced design. Upon approval of the NOI, the applicant was required to file an Application for Certification (AFC), and after approval of the AFC, the applicant was then able to move forward with financing and completion of design and to begin construction of the power-generating facility. Originally the NOI and the AFC processes each took a minimum of nine months; these were consecutive periods and did not include the time required for the review processes and public meetings. The result was the power generators had to allow at least five years before they could start realizing any revenue from their investment. With the advent of the 2001 electrical energy shortages, the NOI and AFC time requirements were reduced. A number of plants were under construction when the CEC regulations were established, and those plants were grand-

[2]Proposed Conference Report No. 1 defines "stranded" costs as the costs of utility power plants and power purchase contracts that cannot be recovered in a competitive generation market. It is applied to nuclear-generating facilities and high-priced long-term contracts.

fathered. The Northern California Power Agency's (NCPA) Geothermal Project No. 2 in Sonoma County was the first power plant fully licensed under the NOI and AFC processes. The author was fortunate to have participated in the design, construction, and startup of this power plant project.

The California Public Utilities Commission (PUC) required the utilities to purchase their energy requirements through the PX on the spot market, that is, without negotiations and at the price of the moment, which can vary. Apparently the utilities had seen the crunch for electrical energy coming and applied to the PUC, seeking permission to enter into long-term contracts for purchase of electrical energy. Southern California Edison applied to the PUC seven times requesting permission to enter into long-term contracts. Each time the PUC refused permission, using the argument that if the long-term contracts were allowed and if an energy glut occurred, the PUC feared the utilities would be locked into high energy prices. In the crisis years, a glut never materialized—to the contrary, a shortage materialized, the demand for electrical energy skyrocketed, and prices zoomed upward. The utilities asked the PUC to allow increases in consumer electrical energy rates sufficient to cover their increased wholesale costs. The PUC, along with Governor Davis, steadfastly refused the rate increases, sending the utilities into enormous debt to the tune of several billions of dollars. It also forced PG&E into the bankruptcy court. The PUC failed in its fiduciary responsibility of permitting the utilities "recovery of their costs, including a reasonable rate of return on their investment."

In the ten-year period from 1990 to 2000, California's electrical energy consumption increased by 1.5 percent with a demand increase of 4 percent; these energy increases were attributed to California's booming economy.

The California population in July 2002, according to the Department of Finance data, was 35.3 million people. California's population in 2025 has been projected to be 52 million. With a current electrical energy maximum demand of approximately 45,000 Mw, the current demand per individual is 1.125 kw. Using that figure, we can expect the electrical demand in 2025 to be 58,500 Mw, an increase of 13,500 Mw or about 587 Mw per year based solely on population increases. Additionally, there should be a component to allow for new construction and technological advances.

The Initiation of the Energy Crisis

The decades-old business structure of electrical-energy monopolies was changed under deregulation. The course of electrical deregulation was directed by Assembly Bill 1890 and Senate Bill 90. Beginning March 31, 1998, many of California's electricity consumers were allowed direct access to choices in their purchases of electricity. However, supposedly because of flaws in the deregulation process, the California Public Utilities Commission eventually suspended direct access in September 2001, much to the chagrin of large consumers.

California was experiencing a ballooning economy, the result of general national economic growth and an expanding population. The state had been relying on its neighbors for approximately 20 percent of its electrical energy. In late 1999 and extending into 2000, those neighboring states found they needed the surplus energy they had been sending to California, and the sale of that energy diminished to its termination. California at that time had had a tacit ten- or fifteen-year moratorium on the construction of new power plants. This combination of circumstances left the state without adequate generation to meet its electrical demand. The energy crisis began in midyear 2000, with the critical months commencing in November 2000 and extending through May 2001. In 2001 there were a total of forty-two hours of blackouts in thirty-one days. The ISO would issue a "Stage 1" alert when available energy reserves fell below 7 percent and scheduled blackouts when the reserves fell to less than 1.5 percent.

Prior to the crisis, wholesale electrical energy prices fluctuated between \$20 and \$50 per megawatt hour (mwh), but by the spring of 2000, wholesale prices exceeded \$100 per mwh. In 1999 the state's wholesale power costs were \$7.4 billion; in the years 2000-2001, that cost escalated to \$27 billion per year. It has been estimated that the crisis cost the state's economy over \$40 billion over a two-year period, and that number does not include the effects of blackouts and the loss of economic growth. In an effort to keep the state's utilities from total financial collapse, the State of California stepped in and embarked on a program of purchasing energy to meet California's electrical energy demand. Starting in January 2001, the State borrowed \$7 billion from its general fund for the purchase of electrical energy, with the pledge of future repayment from an interim loan, followed by the eventual sale of energy bonds. State officials ultimately negotiated \$42 billion of long-term contracts over ten years for the ongoing purchase of electrical energy as a hedge against skyrocketing wholesale prices. The average energy price resulting from those negotiations was $5.5 per mwh.

Considerable blame for the crisis has been passed around by everybody—the blamed included the passage of AB 1890, the shortage of intrastate generation facilities, and limited electrical and natural gas transmission facilities. The crisis was exacerbated by corporate mischief from suppliers. In total, these elements made their contribution to California's energy problem, along with the fact that the state's utilities and regulatory agencies had been undergoing growing pains in their transition from a regulated monopoly to a deregulated, restructured electrical market, all of which provided tremendous challenges for the utilities, regulators, consumers, and politicians.

Bankruptcy

With the skyrocketing wholesale prices and the governor's and the PUC's obsession for maintaining low retail energy prices at the expense of the utilities, the utilities' financial condition was jeopardized. PG&E went into debt to the tune of $13 billion, eventually forcing the company into the bankruptcy court, seeking Chapter 11 protection[3]. SCE was driven to the edge of bankruptcy but was rescued when the PUC allowed SCE to pay off $5 to $6 billion of debt with contributions from the ratepayers' payments; the company's cash on hand was reduced along with dividends.

On September 18, 2003, PG&E started to ease out of bankruptcy protection with the acceptance of a reorganization agreement by all participants, including the bankruptcy judge. The Official Committee of Unsecured Creditors (OCC) and PUC had a plan, which appeared to be creditable; the plan received significant consideration. The OCC/PUC believed the plan would provide stability, repay the creditors in full, and allow PG&E to emerge from bankruptcy with an investment grade profile. A summary of OCC/PUC plan contains the following balance sheet:

---

[3] PG&E went into bankruptcy on April 6, 2001 and came out of bankruptcy on April 12, 2004.

Sources of Funds	$ Millions
Cash on Hand	$3,400
Reinstated Claims	
PC Bonds	545
Preferred Stock	432
New Securities	
Amortizing Bank Loan	1,500
FMB due 2007	1,925
FMB due 2012	1,924
FMB due 2032	1,924
Subordinated Bonds	1,000
New Preferred Securities	500
Total Sources	**$13,150**

Uses of Funds	$ Millions
Class 1 & 2 Priority Claims	$1,354
Class 3 FMBs	2,699
Class 4 PC Bonds	1,613
Class 5 Unsecured Claims	4,570
Class 6 ISO/PX/ Generator Claims	1,601
Class 7 ESP Claims	520
Class 10 Convenience Claims	60
Class 11 QUIDS Claims	301
Class 13 Preferred Stock Claims	432
Total Uses	**$13,150**

Additional liquidity would be provided by $1.9 billion of undrawn revolving credit facilities.

On October 30, 2003, Pacific Gas and Electric Company announced its intention to file a registration statement, early in the following week, with the Securities and Exchange Commission (SEC); the statement included an offer and subsequent sale of up to $9.4 billion of debt securities. The stated purpose of the filing was to initiate the process of registering debt securities that it intended to issue to the public. The filing embraced PG&E's proposed reorganization plan; it was appended to the OCC plan filed in the company's Chapter 11 bankruptcy case. The proceeds of the submitted plan's sales were to be used together with cash on hand and possible other debt financing to pay creditor claims.

The proposed plan of reorganization reflected the terms of the proposed settlement announced in June 2003 between the company and the staff of the PUC. A creditor vote showed overwhelming support for this proposed plan, and on December 17, 2003, the plan received final approval.

California Electrical Energy Supply

California's energy systems depend primarily on five primary fuels: coal, hydro, natural gas, nuclear, and renewables. In 2002 the state produced about 16 percent of the natural gas it used, 53 percent of the petroleum used, and 77 percent of the electricity used. Twenty-three percent of the electrical energy California consumed came from other states. See details in Appendix A.

Data from the U.S. Department of Energy's (DOE) Energy Information Administration (EIA) 1997 report *Primary Energy Consumed in California by Source* indicated California was 3rd in the nation in production of crude oil, 11th in production of natural gas, 3rd in net generation of hydroelectric power, and 6th in nuclear electric energy generation. While it was 2nd in the total amount of energy consumed in the nation, it was 48th in the amount of energy consumed per person. California was 1st in the use of energy in the residential, commercial, and transportation sectors and 3rd in the industrial sector. The state was 2nd in the use of natural gas, petroleum, and electricity. Texas ranked 1st. Appendix B reports PG&E's power content for the years 2002 and 2003.

California has been pursuing new generation technologies and now has 40 percent of the world's geothermal power plants, 20 percent of the installed wind capacity, and 70 – 80 percent of the world's solar electricity generation. Figure 4 shows the distribution of California's generation.

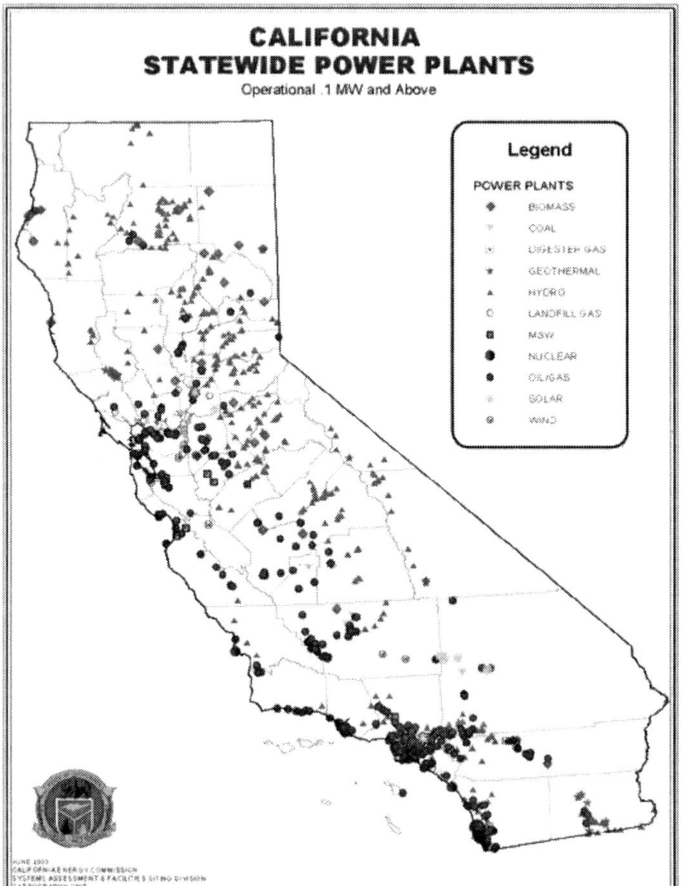

Figure 4.
California Power Plants Statewide
Courtesy of the CEC

Natural Gas Supply

The dominant fuels used in the U.S. economy are oil and natural gas. These fuels provide 62 percent of our nation's energy fuel requirements and almost 100 percent of its transportation fuel requirements. The EIA anticipates the U.S. will need about 50 percent more oil and natural gas by 2020, while the supply is expected to decline over the next two decades. Demand for natural gas is predicted to continue to outpace our domestic production, and the U.S. will be put into the unenviable position of having to increasingly rely on imports of natural gas pipelined from Canada and imports of Liquid Natural Gas (LNG)[4] from producers around the world.

There are some sobering facts regarding natural gas production in the U.S. Domestic natural gas production has peaked, and serious declines in production rates are anticipated. It has been reported there has been only one significant natural gas discovery in the forty-eight states during the first nine months of 2003, and without finding significant new production, supplies will continue to decline in those forty-eight states. New domestic gas production has become increasingly expensive. It is reported that natural gas from the cheapest and most easily obtainable domestic sources has already been found and produced. The remaining gas fields will be more expensive and more difficult to place into production. Even with recent historically high gas prices, drilling and exploration hasn't significantly increased, as is normally expected with high gas prices.

If supply and demand is to be brought into balance, there must be either an increase in supply or a decrease in demand or a combination of the two. If supply does not increase, gas prices will be driven up until gas users are forced to cut back use. In such an event, industries such as chemicals, fertilizer, glass, metals, and others that depend on low energy

---

[4] LNG is simply natural gas that has been super-cooled to -260° F (-158° C). At that temperature, natural gas condenses into a liquid. In liquid form, natural gas takes up to 600 times less space than in its gaseous state, making it feasible for large quantities to be transported over long distances. LNG is odorless, colorless, non-corrosive, and nontoxic, and its weight is less than one-half of water. LNG can be shipped to where it is in demand from many parts of the world where the gas is abundant.

prices could be forced to reduce their production, resulting in a search for dependable low energy prices.

There has been a significant increase in the number of natural gas fired power plants built in the nation in recent years. This has resulted in a 30 percent increase in natural gas demand used for electrical generation, as measured since 1966. Continued construction of new power plants has been estimated to increase natural gas demand by nearly 5 percent per year. Stringent air-quality regulations and new, efficient technologies have made natural gas the fuel of choice for almost all fossil-fuel power plants. Thus, demand for new power plants increases the demand for natural gas. Currently the breakdown of natural gas usage in the U.S. is

- Electrical energy generation: 27 percent
- Industrial: 34 percent
- Commercial: 15 percent
- Residential: 24 percent

For over a decade, natural gas has supplied more than 30 percent of California's total energy requirements. Now, due to the reasonable price of natural gas and its clean-burning characteristics, natural gas has definitely become the fuel of choice within California and is increasingly the fuel of choice for electrical energy generation. With the requirement for more electrical generation, the demand for natural gas has to increase. Almost 84 percent of California's natural gas supplies are obtained from sources outside the state—46 percent from the Southwest, 28 percent from Canada, and 10 percent from the Rocky Mountain area. One-third of the state's electrical energy is generated by gas, and this figure is projected to reach 38 percent by 2009.

During the next two decades, natural gas is anticipated to play a major role in achieving California's environmental objectives. In the last decade, three new interstate gas pipelines were built to service California, augmenting the already over one million miles of existing pipelines that connect the state to gas-producing areas. California's demand for natural gas in 1990 topped 2,025 trillion cubic feet. These new pipelines, operating at full capacity, will provide California with a critically essential additional 2.1 billion cubic feet of natural gas per day. California's pipelines statewide are shown in Figure 5.

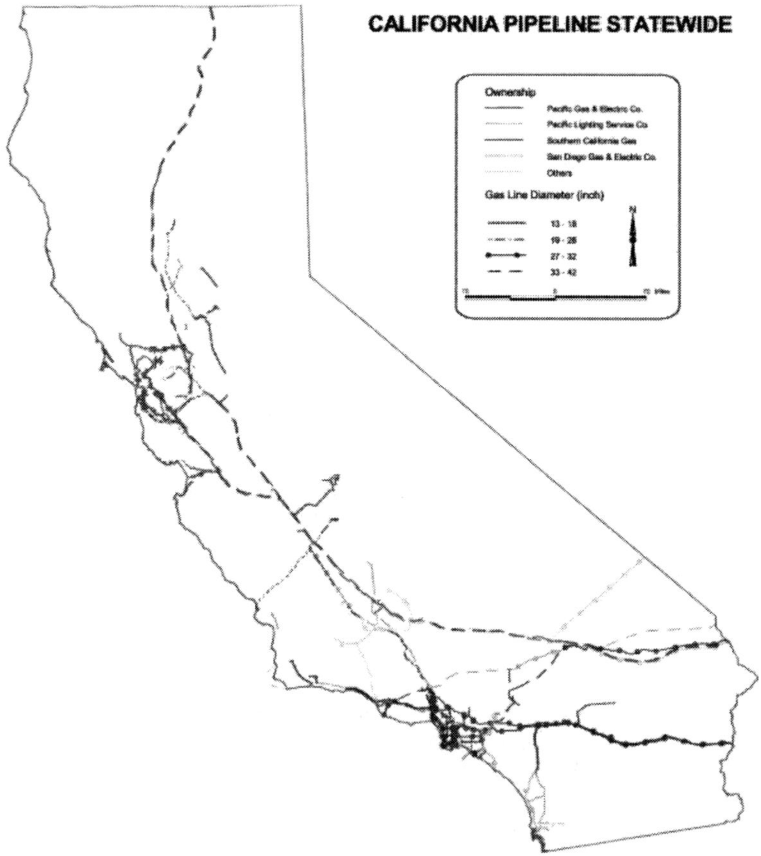

Figure 5.
California Pipelines Statewide
Courtesy CEC

The gap between supply and demand for natural gas may well be bridged with the importation of LNG products. Sempra Energy, parent of SDG&E, is developing two LNG regasification terminals in North America and is anticipated to become a major importer. Importation of LNG will increase our natural gas supply, helping to lower the price of natural gas, which in turn will help the economy. The use of LNG has proven to be safe and reliable in the U.S. since 1944.

II. Twilight—The Crisis's Key Players[5]

There were a number of key participants in California's energy crisis—actually, there were sixteen entities and a number of QFs that participated in the crisis, some to a greater degree than others. Each had specific functions in California's electrical industry, and each specific function was significant. Prior to the crisis, each had specific identifiable areas of responsibilities. With the onset of the crisis, their relationships became completely intertwined, some more than others, but all had an impact on the crisis in varying degrees. What follows is an introduction to each of the entities, outlining its respective normal function followed by an explanation of its role in the crisis.

GOVERNMENT AGENCIES

Federal Energy Regulatory Commission (FERC)

The Federal Energy Regulatory Commission is an independent federal agency responsible for regulating the interstate transmission of electricity, natural gas and oil, and hydroelectric projects. Its responsibility includes (1) regulation of wholesale electricity sales and transmission in interstate

---

[5] The information contained herein has been substantially obtained from the players' respective web sites. Additionally, the information for the ISO, PUC, CEC, Sempra Energy, Duke, and Reliant energy has been substantially supplemented through direct communication with those entities.

commerce, sale of natural gas in interstate commerce, transmission of oil by pipeline in interstate commerce; (2) licensing and inspection of private, municipal, and state hydroelectric projects; (3) approval of the siting of interstate natural gas facilities, including pipelines and overseeing the abandonment of facilities; (4) oversight of environmental matters related to natural gas and hydroelectric projects and major electrical projects; (5) administration of accounting and financial reporting regulations and the conduct of regulated companies; and (6) certification of QFs and general implementation of the Public Utility Act (PURPA). From the FERC web site:

> In 1935 Congress passed the Federal Power Act, which imposed a statuary mandate on the Federal Power Commission, the predecessor to the Federal Energy Regulatory Commission (FERC), to set "just and reasonable" wholesale electricity prices. An accepted standard for just and reasonable [costs] are those that recover production costs, including a "fair" rate of return on the capital invested by the firm. Moreover, if FERC finds that wholesale electricity prices are unjust and unreasonable, the Federal Power Act gives it authority to take actions that result in just and reasonable prices. Finally, the Federal Power Act requires that FERC order refunds for any payments by consumers for prices in excess just and reasonable levels.
>
> Approximately 10 years ago FERC embarked on an explicit policy to promote wholesale electricity markets throughout the U.S. The price a generation owner receives from selling into a wholesale electricity market is determined by the willingness of all generation unit owners to supply electricity, rather than an administrative process that uses the firm's production costs and a rate of return on capital invested to determine the price it receives for electricity.

Included in FERC's promotion of deregulation in the individual states was the requirement that each state establish an Independent Service Operator (ISO) that would be supervised by FERC. The ISO's function is to schedule energy from available power sources to meet load requirements and to schedule the transmission facilities necessary to transport that energy to the scheduled loads and monitor that activity. California's entrance to deregulation occurred with the passage of AB 1890. This bill also established California's PX, a clearinghouse for the purchase of electrical energy. As a result of FERC's supervision of California's ISO, FERC now has implied supervision over California's intrastate transmission facilities.

FERC failed to live up to its responsibility to "produce just and reasonable markets" in the years 2000, 2001, and 2002 by allowing skyrocketing wholesale prices to occur. California and other western states became outraged with the soaring wholesale cost of electrical energy and sought relief from FERC. FERC procrastinated over the wholesale energy prices before finally placing a cap on them. FERC initially blamed California for its electrical energy crisis problems, saying California's deregulation was fatally flawed. The consensus in California was that FERC failed in its responsibility to assure wholesale prices that were "just and reasonable." The investor owned utilities (IOUs) met their responsibility to serve. They paid outrageous prices for the energy they supplied to consumers but were restricted from passing on the high prices to those users. Consequently, one IOU ended up in bankruptcy and another went to the brink.

Public Utilities Commission (PUC)

The California Public Utilities Commission (PUC) was established in 1911 through a California Constitutional Amendment titled the Railroad Commission. In 1912, the legislature passed the Public Utilities Act, expanding the PUC's regulatory authority to include natural gas, electricity, telecommunications, water companies, railroads, and marine transportation companies. In 1946 the Commission was renamed the California Public Utilities Commission. The PUC has the responsibility of assuring that California utility customers have safe, reliable utility service at reasonable rates. Its jurisdiction includes fraud protection.

The Commission establishes service standards and safety rules, and it plans and authorizes utility rate changes. The PUC monitors safety and oversees markets to constrain any anticompetitive activity. The PUC has prosecutorial authority for unlawful utility retail marketing and billing activities, including business relationships between utilities and their affiliates. The PUC receives and resolves customer complaints against utilities. It designs and implements energy-efficiency programs and sanctions low-income rates for disabled customers. The PUC has the responsibility to superintend the merger and/or restructuring of utility corporations, and it enforces the California Environmental Quality Act for construction of utility facilities. The PUC coordinates with other state and federal agencies in the promotion of water quality, environmental protection, and safety. It will also intervene in federal proceedings on issues that impact California utility rates and/or services.

The PUC has five commissioners appointed by the governor and confirmed by the state senate; each appointment is for staggered six-year terms. The governor selects one of the five to serve as commission president. The PUC employs economists, engineers, administrative law judges, accountants, lawyers, and safety and transportation specialists. The staff is organized into several units, including an advisory unit, an enforcement division, and a strategic planning group. The Office of Ratepayer Advocates, an independent arm of the PUC, represents consumers in commission proceedings. The PUC employs a public advisor to assist the public in participating in commission proceedings; it also has a unit charged with informally resolving consumer complaints.

Prior to the initiation of a competitive energy market under AB 1890, private utilities operated within a PUC-regulated market. The PUC supervised the utilities' costs of generation, transmission, and distribution. It allowed the utilities to recover their costs and earn a reasonable return on their investment. Under AB 1890, the PUC lost regulation of generated power, except for that which was retained by the utilities. Wholesale-generated power by marketers then came under the purview of FERC. The PUC regulations, as a result of AB 1890, required the utilities to sell the majority of their fossil-fuel generating plants and denied them opportunity to enter into long-term energy purchase contracts—the objective being to prevent the utilities from dominating the energy market. Eventually the PUC did allow the utilities to enter long-term contracts up to one-third of their minimum load. Subsequently in 2000, the PUC allowed expanded purchases, but reserved its right to disallow any cost above the spot market price. In other words, if the spot market price fell below the utilities' contract prices, the utilities were responsible for the difference.

AB 1890 allowed direct access, which permitted consumers to negotiate directly with the generator or marketer of electrical energy for their energy requirements. Many large consumers exercised this option when spot market energy prices were high. This option was not popular with residential customers and small businesses. Those suppliers that did enjoy residential and small businesses customers were generally "green-energy" (energy from QFs) suppliers. The green-energy suppliers appeared to be the only suppliers that advertised the availability of their electrical energy. The large consumers returned to the utilities when the wholesale prices soared. On January 1, 2001, the Department of Water Resources began purchasing power for the utilities and eventually entered into long-term energy contracts with electrical energy suppliers. State officials then realized that if the wholesale price of energy fell below the contract prices and if the consumers returned to direct access, they would bypass paying for the DWR energy purchases, leaving the remaining consumers with the obligation for payment of the long-term contract energy. The PUC felt that in fairness, direct access should be terminated so that all consumers would pay for the long-term contracts. Subsequently, direct access was again allowed; however, those that did return to direct access were charged an "exit fee" which was applied to the DWR's long-term contract costs. The exit fee was designed to include the direct-access buyers in the repayment of energy purchased by the DWR in its long-term contracts, i.e., avoid giving them a free ride.

Independent System Operator (ISO)

In 1996 the California legislature passed AB 1890, committing the State to restructure the electrical energy industry. The law included the establishment of the California Independent Service Operator (ISO). The ISO's assigned mission was, and is, to serve the consumers in California in the following ways: (1) planning and operating a safe and reliable electrical system, (2) managing open and nondiscriminatory electrical transmission facilities and services, (3) facilitating appropriate investment in electrical transmission and generation infrastructures, and (4) overseeing efficient wholesale electrical markets at minimum cost.

Since the FERC regulates the ISO, the ISO is required to adhere to FERC tariffs and other FERC rulings as applicable. However, the ISO's board of governors is appointed by the governor of the state of California. While there is no direct chain of command, similar to other state agencies like the Department of Motor Vehicles, the board structure ensures that the ISO is committed to doing what's best for the people of California.

AB 1890 required California's three IOUs (Pacific Gas & Electric, Southern California Edison, and San Diego Gas & Electric) to turn over the management of their transmission facilities to the ISO. The IOUs retain ownership, physical operation, and maintenance of those facilities, while the ISO manages the use of the transmission systems and ensures fair and open access to the transmission capacity.

In its operation, the ISO accepts scheduled transmission of electrical energy in a day-ahead time frame from entities wanting to use the transmission grid. It is imperative that the ISO see the energy schedules, even if it is not a party to the transaction, in order to make sure that all the scheduled energy will "fit" into the available transmission capacity. If it does not, the ISO will allocate available capacity from the ISO's congestion management market and if need be, obtain energy from other sources to make up for prescheduled energy that may not be accommodated on the grid. For example, say on a Monday, the ISO receives schedules from generators and utilities that indicate how much energy is to flow on Tuesday and from which generator and to what load.

As a result of the energy crisis's impact on the IOUs financial condition, the Department of Water Resources (DWR) was assigned the task of purchasing energy for the cash-impoverished IOUs in order to maintain them as viable entities. The purchases were long-term contracts financed through bonds. Consequently, much of the energy was delivered under Department of Water Resources contracts. The ISO was not a party to those financial arrangements, and those arrangements had no impact on the ISO's real-time activities to balance the grid. The IOUs use the power from those contracts as part of their portfolios, in combination with their retained generation.

The ISO had no policing authority over market-power abuse during the energy crisis months. It did not have any direct ability to control market power. The ISO's Department of Market Analysis (DMA) identified abuses and was to report them to FERC and others but had limited ability to react. Now the ISO enforces the FERC West-Wide Market Power mitigation protocols. FERC gave the ISO the authority to analyze bidding behavior, and if it's apparent that market power is being exercised, it can submit proxy bids (lower bids) for power that is being offered at outrageously high prices. There was also in place a "must offer" requirement; meaning generators could not withhold power from the market place.

Department of Water Resources (DWR)

DWR operates and maintains the State Water Projects, including the California Aqueduct. The department also provides dam safety and flood control services, assists local water districts in water management and conservation activities, promotes recreational opportunities, and plans for future statewide water needs.

On January 17, 2001, the DWR was assigned the task of purchasing electrical energy for California's three major IOUs[6] to supply energy for more than 27 million people during the crisis period when supplies were scarce and prices were skyrocketing out of control. It was an ominous time in California history, both figuratively and literally. The California ISO relied on rotating blackouts to prevent a widespread collapse of the state's electrical grid, and the major utilities were in deep financial trouble. Meanwhile, energy supply companies took full advantage of a dysfunctional energy market.

Within hours of the governor issuing an Emergency Declaration, DWR mobilized and created a new division within itself—the California Energy Resources Scheduling (CERS) division. Staffed with state employees and hired consultants, CERS gathered in a hastily-prepared emergency operations center in Sacramento to begin daily power purchases. The CERS trading room was a twenty-four-hours-a-day, seven-days-a-week operation where high-stakes bartering took place with sellers who knew that California was desperate for electricity. Average costs for energy were peaking at more than $300 per mwh, nearly ten times the fair market value. With a loan from the State's General Fund, CERS spent more than $1 billion a month in the first half of 2001. CERS staff worked round the clock. During the summer of 2001, CERS purchased 17,000 Mw of power, enough electricity to supply 17 million homes and businesses.

To do that, CERS began to reduce the amount of electricity it needed to buy out of the volatile daily spot market by negotiating long-term contracts that would lock in more stable, predictable prices. The contracts

---

[6] Pacific Gas & Electric (PG&E), Southern California Edison (SCE), and San Diego Gas & Electric (SDG&E)

had an additional benefit desperately needed in California—they brought more than 5,000 Mw of newly-built power facilities to the state. In total, DWR entered into fifty-six long-term agreements in 2001 at a cost of $42 billion over twelve years. In October and November 2002, CERS sold $11.2 billion in bonds—the largest bond sale in California's history—to pay for its power purchases. Bond proceeds were used to repay $6.5 billion to the General Fund, retire a $3.4 billion short-term loan, and to maintain adequate reserve levels.

CERS concluded its daily power purchases operation on January 1, 2003, successfully returning that daily responsibility to the utilities. Currently, the CERS continues to manage its long-term contract portfolio as well as its bond debt. The task of effectively managing the State's portfolio of contracts and bond financing is important for maintaining reliable energy supplies at the lowest possible cost for California ratepayers. Both of these complex and critical responsibilities will continue until the last of the contracts expires in 2015 and the bond is completely retired in about 2020.

California Energy Commission[7] (CEC)

The Energy Resources Conservation and Development Commission, now known as the California Energy Commission (CEC), was established by the legislature in 1974 to address the energy challenges that resulted from OPEC's attempted price embargo in the early 1970s. The Warren-Alquist State Energy Resources Conservation and Development Act (usually referred to as the Warren-Alquist Act) created the CEC, and it was signed into law by then Governor Ronald Reagan. The Energy Commission is the state's principal energy policy and planning agency. The measure assigned the CEC five major responsibilities: (1) forecasting future energy requirements and recording historical energy data, (2) licensing thermal energy power plants rated 50 Mw or larger, (3) promoting energy efficiency through the creation of appliance and building standards, (4) developing energy technologies and supporting the implementation of renewable energy, and (5) planning for and directing the state's responses to energy emergencies.

---

[7] The CEC's role during the crisis was primarily limited to the expediting of power plant licensing.

The CEC has five members appointed by the governor, each to staggered five-year terms, with a chairman and vice chairman selected from among the members for two-year terms. The commission appointments require senate approval. By law, one member must be selected from the public at large, with the remaining members representing the fields of engineering and/or physical science, economics, environmental protection, and law. The CEC nominates a public adviser, confirmed by the governor, for a term of three years. The public adviser has the responsibility for assuring that the public and all other interested parties are adequately represented at all CEC proceedings. The CEC's Executive Office provides the staff and management support to the CEC and directs the commission's overall policy and management.

The CEC's General Counsel's Office staff attorneys represent the CEC in administrative hearings before the PUC and FERC, in intrastate and federal court proceedings, and in internal proceedings such as power plant siting cases. CEC hearing advisers serve as administrative law judges for the CEC on power plant siting and licensing cases.

The CEC forecasts future energy trends through various reports. Publications are available through the CEC's Publications Unit and are available in electronic form on the CEC's web site, www.energy.ca.org.

The CEC is funded through an electrical consumption surcharge collected in customer rates of the investor-owned utilities and transferred to the state treasury. The surcharge is .0002¢ per kwh. The CEC's budget is supplemented by federal money for specific energy-efficiency programs. The CEC also manages any public-good surcharges used for energy research and development and renewable energy programs.

With the passage of AB 1890, the CEC's role was expanded to include overseeing funding programs supporting public interest energy research; advancing energy science and technology through research, development, and demonstration; and providing market support to existing, new, and emerging renewable technologies.

An energy consortium of California's principal energy agencies—the California Energy Commission, the Consumer Power and Conservation Financing Authority (CPA), and the Public Utilities Commission—designed an Energy Action Plan that identifies specific goals and actions to avoid energy outages and excessive price spikes in electricity or natural gas. The consortium hoped the initiatives would send a signal to the market that California really was a good place to do business and that investment in more efficient use of energy, in new electricity, and in natural gas infrastructure would be rewarded. The consortium recognizes that California currently has a hybrid energy market. They believe the revised state policies with positive regulation will promote the best features of a vigorous, competitive wholesale energy market. At the same time, they remain mindful of the necessity to maintain energy rates that are affordable, and they attempt to be generally sensitive to the implications of energy policy relative to global climate change and the environment.

This energy consortium intends to achieve its objectives by (1) meeting California's energy growth requirements, including optimizing energy conservation and resource efficiency and reducing per capita electrical demand; (2) ensuring reliable, affordable, and high quality power supply by constructing adequate new generation; (3) accelerating the state's renewable resource generation goal by 2010; (4) upgrading and expanding the electrical transmission and distribution infrastructures and expediting activation of the necessary facilities; (5) promoting customer and utility distributed generation ownership; and (6) ensuring reliable natural gas supplies at reasonable prices.

The consortium recognizes that the State has to guide development of the state's energy system in the public's long-term interest. It has to anticipate potential problems, and it has to make timely decisions in resolving problems. The consortium specifically commits to (1) providing decision-makers with impartial assessments of the state's current and future electrical and natural gas demands, resources, and prices; (2) licensing and, if necessary, funding construction of new energy facilities that comply with the reliability, economic, public health, and environmental requirements; (3) assuring that the utilities will be able to carry out their obligations to serve, including the maintenance of adequate reserves; (4) restoring private sector confidence in California's energy markets to encourage investment in California's economy; (5) developing an "early warning" system that will alert policy makers to potential future problems; (6) cooperating with FERC to redesign market rules and prevent manipulation of the energy markets; (7) partnering with governmental and other groups in western North America to pursue mutually-held energy goals; and (8) continuing California's progress in achieving the state's environmental goals and standards, including minimizing the energy sector's impact on climate change.

Air Resources Board[8] (ARB)

The California Air Resources Board (ARB) is the state agency responsible for providing the California population with a safe, clean, healthy atmosphere through the reduction of emissions from motor vehicles, fuels, consumer products, and sources of air toxins at the state level. Additionally the ARB oversees local air pollution agencies.

Air quality has improved substantially over the thirty years since the ARB was formed; the ARB believes additional work remains to be done. The ARB believes the public has the opportunity to participate in the continuing development of air pollution policies, plans, and regulations. The ARB believes the best way to ensure public input and environmental justice is for the public to have an understanding of the process for air quality decision-making. County or regional air pollution control districts and the air quality management's thirty-five local air districts are responsible for developing programs to improve air quality through local attainment of air quality standards. These local air districts are generally the first point of contact for resolving local air pollution issues and/or complaints. Each local air district regulates stationary sources of air pollutants, such as industrial and commercial facilities, power plants, construction activities, outdoor burning, and other non-mobile sources of pollution; the responsibility includes enforcing air pollution rules and regulations at the local level.

California's Air Resources Board is composed of eleven appointees, which include five local elected officials, and has the following responsibilities: (1) regulates mobile sources (such as cars, trucks, and buses), fuels, consumer products, and air toxins at the state level; (2) oversees and assists local air districts; and (3) sets state ambient air quality standards.

The U.S. Environmental Protection Agency (EPA) has national responsibilities as follows: (1) sets federal ambient air quality standards; (2) oversees and approves state programs; and (3) regulates at the national

[8] The ARB's role during the crisis was limited to the relaxation of some air-quality standards for given periods of time.

level pollution that arises from cars and trucks, trains, aircraft, and most construction and farm equipment.

The California Health and Safety Code (H&SC) section 39607(e) requires the Air Resources Board to establish area designation criteria and periodically review those criteria. The designation criteria provide the basis for the board to designate areas of the state as "attainment," "nonattainment," or "unclassified" for the state standards. H&SC section 39608 requires the board to use the designation criteria to label areas in California relative to nine pollutants: ozone, sulfates, carbon monoxide, lead, nitrogen dioxide, hydrogen sulfide, sulfur dioxide, visibility-reducing particles, and suspended particulate matter (PM10). Each year, the board is to review the area designations and update them as appropriate, based on the three most recent calendar years of air quality data. For example, the board made area designations in 1999 based on the data collected from 1996 through 1998.

The Air Resources Board has the responsibility for conducting rule makings to adopt and amend regulations covering a variety of areas. If a company plans to construct and operate an electrical power plant, it is required to purchase "air credits" and apply to the local Air Resources District in which the plant is to be located. The air credits can be purchased from other companies that have reduced their emissions by the installation of emission controls or have permanently shut down their operations.[9]

INVESTOR-OWNED UTILITIES (IOU)

Pacific Gas & Electric (PG&E)

Pacific Gas and Electric Company has become one of the largest combination natural gas and electric utilities in the U.S. The company, a wholly owned subsidiary of PG&E Corporation, is based in San Francisco. The company has 19,575 employees carrying out Pacific Gas and Electric Company's primary business—the generation, transmission, and

---

[9] In a letter to the El Dorado County ARB, the author requested information on how the "air credits" were determined; no response was received.

delivery of electrical and natural gas energy to approximately 13 million customers throughout its 70,000-square-mile service area of Northern and Central California. Deregulation required PG&E to divest itself of many of its fossil-fuel power plants. The California Public Utilities Commission regulates the Pacific Gas and Electric Company. PG&E's service area stretches from Eureka in the north to Bakersfield in the south, and from the Pacific Ocean in the west to the Sierra Nevada in the east. Its facilities include 136,500 circuit miles of electric lines, 45,000 miles of natural gas pipelines, 4.8 million electric customer accounts, and 4.0 million gas customer accounts.

Southern California Edison (SCE)

Southern California Edison provides electrical energy for 11 million individuals, 427 communities and cities, 5,000 large businesses, and 280,000 small businesses in Southern and Central California. This service area has 16 utility interconnections, 4,900 transmission and distribution circuits, 365 transmission and distribution crews, 12,642 employees, and over a century of experience. SCE says its heritage is the basis for its reliable electrical service. It reports that for more than one hundred years, SCE has provided high-quality, reliable electrical service for more than 4.2 million business and residential customers over a service area of 50,000 square miles in Coastal, Central, and Southern California. SCE says it is aware that the electrical energy industry is changing rapidly and dramatically. Global competition dictates the way companies will have to do business—energy issues are increasingly becoming worldwide. At the same time, SCE says providing reliable local service has grown progressively more important.

California Assembly Bill 1890 placed utilities, including SCE, in a deregulation market, a new era for the state's electrical utility industry. In 1997, as part of restructuring the electrical industry in the state, SCE sold its twelve fossil-fuel generating stations and reorganized nearly every aspect of its business to prepare for the changing competitive market environment. SCE still owns and operates its hydroelectric and nuclear power facilities serving its area, but SCE's main business role is now primarily converted to that of providing power transmission and distribution services. SCE is a PUC-regulated public utility.

San Diego Gas & Electric (SDG&E)

SDG&E is a regulated public utility providing energy service to three million consumers through 1.3 million electric meters and 789,000 natural gas meters in San Diego and southern Orange counties. SDG&E is part of Sempra Energy Utilities, which is the umbrella for Sempra Energy's regulated California utilities. Sempra Energy is based in San Diego and is a Fortune 500 energy services holding company. SDG&E is a PUC-regulated public utility. Southern California Gas Company (The Gas Company) is a member of Sempra Energy Utilities and is the nation's largest natural gas distributor, serving 18 million consumers through 5.1 million meters. The company's service territory encompasses 23,000 square miles, mostly in Central and Southern California. SDG&E and The Gas Company are Sempra Energy's only regulated business units.

On October 8, 2003, Sempra Energy proposed a plan to buy a mix of power purchases and local generation assets, including energy from renewable resources and new power plants, to ensure a safe, reliable supply of energy to its service region. The proposed plan reflected Sempra Energy's response to the State's decision, which became effective Jan. 1, 2003, to put the three major utilities, including SDG&E, back into the role of planning and procuring electricity for their customers, replacing the DWR purchasing activities that had resulted from the energy crisis of 2000-2001. SDG&E's plan is subject to PUC approval requiring public hearings, which are expected to be held by the end of 2003, with a final decision anticipated in the first quarter of 2004.

SUPPLIERS

Enron Corporation (Enron)

Enron Corporation was one of the world's leading energy, commodities, and services companies. Enron had revenues of $101 billion in 2000, marketing electrical energy and natural gas, delivering energy and other physical commodities, and providing financial and risk management services to its customers around the world.

Enron appears to be a holder of considerable business ventures. To determine which of those ventures were involved with California's energy crisis has been difficult. The one venture that is described to some extent is Enronwind.

Enron considers Enronwind a pioneer and leader in the wind power generation market. Enronwind provides a wide array of power plant services, including project design and engineering, project development, financing, and operating and maintenance services.

Many nations are turning to alternative energy resources to help meet the escalating demand for electrical energy. Enron contends that wind power will be in the forefront. Wind is the least expensive renewable energy resource and is cheaper than coal, oil, nuclear, and most natural gas fired generation. With an energy cost of approximately 3.5 – 4¢ per kwh and declining, wind energy is becoming more attractive to utilities and electric cooperatives. Wind power is approaching 50,000 Mw worldwide.

Calpine Corporation (Calpine)

As a leader in independent power production, Calpine advertises itself as "repowering America" to meet the needs of an economy that is demanding cleaner sources of electrical energy. Calpine is the leading producer of geothermal energy in the U.S. and the world. Its programs of developing and building modern natural gas fired electrical energy-generating facilities are the largest in the electrical industry's history. Calpine has accumulated nearly ninety electrical energy-generating plants that combine the energy resources of natural gas fired and geothermal power. Its total production capacity exceeds 22,000 Mw.

Duke Energy (Duke)

Duke Energy has been in the energy business for nearly a century and anticipates growing in the vital marketplace of world energy. Duke is one of the largest investor-owned utilities in the U.S. It advertises itself as the largest producer of natural gas in the U.S. and is among the top ten electrical generation marketers. It is a wholesale energy marketer providing competitively priced, reliable electrical and natural gas energy to millions of consumers. It services a 22,000 square-mile area of the Carolinas, where it serves approximately 2 million retail customers. Duke operates through 12,700 miles of transmission lines, 66,200 miles of distribution lines, and a vast system of transmission towers, utility poles, transformers, insulators, and conductors. Duke also builds electrical generation and natural gas infrastructures. It designs, builds, owns, and operates electric-generating facilities and participates in the management and trading of reliable energy. Duke Energy owns and operates peaking and combined-cycle power plants in key regions across North America. Its peaking generating facilities are operated during times of increased demands over base load requirements. Duke-operated plants in the western states are listed below.

Location	Facility	Mw	Ownership
California			
Oakland	Oakland Power Plant	165	100%
Monterey Company	Moss Landing Power Plant	2,538	100%
Morro Bay	Morro Bay Power Plant	1,002	100%
Chula Vista	South Bay Power Plant	706	10-year lease
Arizona			
Mohave Company	Griffith Energy Facility	600	50%
Maricopa	Arlington Valley Energy Facility	570	100%
Washington			
Grays Harbor	Grays Harbor Energy Facility	650	100%
Nevada			
Clark Company	Moapa Energy Facility	1,200	100%
New Mexico			
Luna Company	Deming Energy Facility	570	100%
Canada			
Taylor BC	McMahon Cogeneration Plant	116	50%
Total		**8,117**	

NRG Energy/Dynergy

On November 20, 2000, NRG Energy and Dynergy announced a major expansion of their partnership portfolio in the western states with their purchase of 1,330 Mw of power-generation facilities from Sierra Pacific Resources. The facilities included the Clark and the Reid Gardner Generating Stations, which serve the growing Las Vegas market. Included in the agreements was a power purchase agreement with Nevada Power for their purchase of energy and ancillary services until March 1, 2003. The sale of the acquired generation assets was the result of a regulatory precondition in a Sierra Pacific-Nevada Power merger, which was completed in July 1999. The Reid Gardner and Clark Stations were among seven facilities included in a Sierra Pacific auction. NRG was also the successful bidder for Sierra Pacific's 50 percent interest in the 522 Mw coal-fired North Valmy Generating Station. Additionally, NRG acquired 100 percent interest in 25 Mw of peaking units in northern Nevada, the partnership's sixth facility, which expanded its California capacity to a total capacity of 2,768 Mw of power generation. In combination with other assets developed and acquired throughout the nation in the year 2000, the addition of these facilities to the partnership's national generation portfolio increased the partnership's generating capacity by more than 4,000 Mw. This placed the partnership in a competitive position with many other industry mergers without the regulated transmission and distribution.

NRG operates the stations while Dynergy is the power marketer and fuel supplier. The Clark Generating Station, located in southeastern Las Vegas, consists of ten natural gas and oil-fired generating units; with a total capacity of 740 Mw. Natural gas is the primary fuel, with fuel oil used as a backup. The Reid Gardner Generating Station, located 52 miles northeast of Las Vegas, consists of four base-loaded coal-fired units. Three of the units, wholly owned by Nevada Power and California's DWR, are each rated 110 Mw. The NRG and Dynergy partnership acquired Nevada Power's interest of 260 Mw in the fourth 275 Mw coal-fired unit as part of the transaction; California's DWR retained its 15 Mw ownership interest in the unit.

NRG is a wholly owned, unregulated subsidiary of Xcel Energy, which has a diverse electrical power generation portfolio around the globe with a variety of fuel types and facilities. NRG generation facilities can be found throughout North America, Europe, Asia-Pacific, and Latin America. NRG projects are designed to use diverse fossil fuel sources, including natural gas, oil, coal and coal seam methane, and green-power renewable fuels, including biomass, landfill gas, hydro, geothermal energy, and refuse-derived fuels. Western state power plants owned by NRG are listed in the following table.

Location	Facility	Fuel	Mw	Ownership
California				
El Segundo	El Segundo	Natural gas	510	50%
Carlsbad	Encino Power Station	Natural gas	482.5	50%
Long Beach	Long Beach Generating	Natural gas	265	50%
San Diego	San Diego Turbines	Natural gas/ diesel	126.5	50%
Nevada				
Henderson	Saguaro	Natural gas/oil	52.5	50%
Total			**1,436.5**	

Dynergy provides electrical and natural gas energy and liquid natural gas to wholesale customers in the U.S. and to retail customers in the state of Illinois. The company owns and operates a diverse portfolio of energy facilities, including approximately 40,000 miles of electric transmission and distribution lines and power plants with a net generating capacity of approximately 13,000 Mw.

Dynergy's gas processing plants can process more than two billion cubic feet of natural gas per day. Dynergy's LNG business is engaged in the gathering and processing of natural gas and the fractionation, storage, transportation, distribution, and marketing of LNG. Its natural gas gathering and processing operations are located in the high-growth oil and gas exploration and production areas of north Texas and the Louisiana Gulf Coast, and in the mature areas of Permian Basin of west Texas and southeast New Mexico. Dynergy's LNG assets are located in Mont Belvieu, Texas, the hub of the U.S. LNG business, and in Lake Charles, Louisiana. Additionally, it manages a diverse portfolio of storage and transportation assets.

Mirant

Mirant, headquartered in Atlanta, Georgia, is a competitive energy company producing and marketing electrical energy in North America, the Caribbean, and the Philippines. Mirant, as an asset-based company and as a competitive energy provider, is rated among the top producers of unregulated and traditional regulated utility electrical energy generation. It owns and/or controls more than 22,000 Mw of generating capacity and currently has approximately 1,000 Mw under construction. Mirant's generating facilities produce, sell, and deliver reliable energy products and services to utilities, producers, municipal systems, aggregators, electric-cooperative utilities, generators, marketers, and large industrial customers. About half of its assets have the capability to switch among natural gas, oil, and coal. This fuel diversity of the generating plants gives Mirant the flexibility to respond to fluctuating fuel prices and to adjust its generating facilities to meet market power swing demands. Approximately 92 percent of its facilities are located within the proximity of large metropolitan areas. Mirant has facilities near San Francisco, Washington DC, Boston, New York City, Dallas, and Las Vegas. Mirant's western facilities are listed below. Mw figures represent Mirant's net ownership interest, which in some cases may be less than 100 percent of total generation capacity.

Location	Facility	Fuel	Mw
California[10]			
Antioch	Contra Costa Generating Plant	Natural gas	674
Pittsburg	Pittsburg Generating Plant	Natural gas	1,906
San Francisco	Potrero Generating Plant	Natural gas/ distillate	362[11]
Nevada			
Las Vegas	Apex Generating Plant	Natural gas	430
Oregon			
Boardman	Coyote Springs Generating Plant	Natural gas	280
Total			**3,652**

[10]California capacity is 2,942 Mw; Mirant's ownership is 100 percent.

[11]Consists of 205 Mw of base load capacity and 156 Mw of peaking capacity.

Mirant acquired its Pittsburg, Contra Costa, and Potrero generating facilities in April of 1999; since then, Mirant has implemented upgrades to increase electrical output and improve air quality. Additionally, attention is given to the protection of local aquatic life and other endangered species. The three generating plants in the San Francisco Bay Area provide approximately 3,000 Mw of electrical generating capacity to the San Francisco, Northern California, and Central California power grid. Mirant's primary customer for electrical energy is PG&E.

Reliant Resources (Reliant)

Reliant Resources, based in Houston, Texas, provides electricity and energy services to retail and wholesale customers in the U.S., marketing those services under the Reliant Energy brand name. Reliant acquired its first power-generation facilities in April 1998 and has grown to a capacity of about 20,000 Mw in operation, under construction, or under contract in the U.S. The company also trades and markets electrical energy, using its 116 owned power plants in the U.S. to provide reliable supplies of electrical energy to end-use customers through power-purchase agreements and direct sales to utilities to serve their native loads. The company provides a complete suite of energy products and services to approximately 1.7 million electrical customers in Texas, ranging from residences and small businesses to large commercial, industrial, and institutional customers. Reliant also serves large commercial and industrial clients in the PJM (Pennsylvania, New Jersey, and Maryland) Interconnection. Reliant's Energy Wholesale Group is a leading provider of power and has plant operations in twelve states in the following U.S. power regions: PJM, Mid-Atlantic, Northeast, Southeast, Midwest, and West. In the PJM and Mid-Atlantic region, Reliant principally serves customers in Pennsylvania, Maryland, New Jersey, and West Virginia. In the far Northeast, it serves customers in New York, with major plants located in New York City and seventy-one hydro facilities in upstate New York, totaling 672 Mw. It serves the Southeast with plants in Florida, Texas, and Mississippi. It serves the Midwest with plants in Illinois and Ohio, and in the West the states of California and Nevada. In the states of California and Nevada, Reliant Energy owns six electrical power generation facilities. Reliant purchased the Coolwater, Etiwanda, Ellwood, Mandalay, and Ormond Beach plants from Southern California Edison Company in two transactions in April 1998 and July 1998. The following table provides a list of Reliant's current generating capability.

U.S. Region	Operating Megawatts	Under Construction	Renewable Energy	Total Megawatts
Northeast	9,240	1,120		10,360
Midwest	2,687			2,687
Southeast	1,133	804		1,937
West	4,635	548		5,183
Texas	781		260	781
Totals	**18,476**	**2,472**	**260**	**20,948**

Williams/AES (Williams)

Williams currently owns and/or retains an interest in eighty-eight power-generating facilities with a total of over 23,000 Mw in the U.S., Canada, Australia, Argentina, Brazil, Dominican Republic, Pakistan, the Netherlands, Hungary, Kazakhstan, China, and the United Kingdom. Its primary service areas are concentrated in Venezuela, Wyoming, the San Juan Basin, and the Gulf of Mexico. Williams produces, gathers, processes, and transports natural gas for home heating and electrical power generation throughout the U.S. Its key services include interstate natural gas transportation, natural gas exploration and production, and natural gas gathering and processing. Williams's natural gas networks include seventeen gas processing and treating plants with a total capacity of 5.3 billion cubic feet, 8,500 miles of natural gas gathering lines, and innovative deepwater infrastructure. Williams delivers clean-burning natural gas to millions of American consumers via their local utilities and to generators of electrical energy through more than 15,000 miles of its interstate natural gas pipelines—Transco, Northwest, and Gulfstream (partially owned by Williams). Its gas pipeline transportation capacity is more than 12 billion cubic feet per day, enough gas to heat 30 million homes on a winter day, representing approximately 12 percent of the natural gas consumed in the U.S. Williams markets approximately 550 billion BTUs of natural gas production and purchases approximately 700 billion BTUs of fuel for its natural gas processing plants. Approximately 450 million cubic feet per day is the company's total average net production. Williams's U.S. seasonal storage capacity is over 230 billion cubic feet, and its total equivalent natural gas reserves were 2.4 trillion cubic feet at year-end 2002; 99 percent of its reserves are natural gas.

In the West, Williams is developing rich reserves of natural gas, using innovative methods to minimize any environmental impact; additionally, it is working with landowners, farmers, and ranchers to coordinate with them on production schedules and operational issues. Williams's exploration and production activities include the Piceance, San Juan, Powder River, and Arkoma production basins; midstream gas processing plants; separation/dehydration/treating plants; thousands of miles of gas gathering lines; and deepwater infrastructure. It has the capacities to gather, process, and treat billions of cubic feet of natural gas per day and large-scale capacity to produce LNG.

In the late 1990s, Williams established a major presence in California's newly competitive wholesale electricity market. On May 18, 1998, Williams reached an agreement with AES Corporation to supply fuel to market 3,954 Mw of power from fourteen generating units located at three AES electrical generator facilities in the heart of the Los Angeles Basin. The agreement followed close behind Williams's merger with Mapping and Positioning (MAPCO). With the AES transaction, Williams has strengthened its position in the rapidly deregulating energy market. Williams's general manager of marketing and trading business said, "California has taken positive steps to open their markets to competition with the formation of the Independent System Operator and the Power Exchange marketplace. The Los Angeles Basin is an important segment in the Western System Power Pool, an active marketplace, which will enable Williams to effectively manage the risks and profitability of this transaction." Williams retains the right to sell all of the energy and ancillary services produced from the AES electrical generating plants in the Southern California cities of Long Beach (2,083 Mw), Huntington Beach (563 Mw), and Redondo Beach (1,310 Mw). AES has the responsibility to operate the facilities it acquired to ensure operational standards. Williams supplies portions of the fuel via its Kern River interstate natural gas pipeline system.

Qualifying Facilities (QFs)

Qualifying Facilities (QFs) are individuals or entities that produce energy either through renewable power production or cogeneration facilities that are qualified under Section 201 of the federal Public Utility Regulatory Policies Act of 1978 (PURPA). The primary business of QFs is not the sale of electrical energy; rather, they own and/or operate generation facilities. PURPA, among other things, required utilities to pay favorable power rates to two groups of non-utilities: (1) small power producers using renewable energy sources and (2) cogenerators. PURPA permitted these two classes of operations to be designated as qualifying facilities (QFs) under certain conditions. To qualify for QF status under PURPA, both cogenerators' and small power producers' ownership in electric utilities must be less than 50 percent. QF cogenerators under PURPA must produce a form of useful thermal energy in their operations, with the waste heat used for the production of electrical energy, and they must meet certain operating and efficiency criteria. Small power- producing QFs must generally be rated less than 80 Mw, and at least 75 percent of the total energy input must be provided by renewable energy. It is important to note the fact that in the analysis of pricing purchased power; QF cogenerators are not required to use renewable fuels. The utilities are required by PURPA to buy electricity from QFs at rates not to exceed a utility's "avoided cost," i.e., power produced for less than what it would have cost the utility to generate the same power, or the incremental cost to the electric utility of alternative electric energy which the utility would have generated or purchased from another source. FERC is responsible for certifying QFs and general implementation of PURPA, leaving the determination of the utilities' avoided cost to the individual states and their respective utility commissions.

States that had progressive renewable energy policies, such as California's renewable tax credit, influenced the development of renewable energy technologies. PURPA remains the major promoter behind the massive growth in the number of non-utility power producers. After an initially rapid expansion, the number of new filings for QF status decreased as the cost of alternative energy sources, which formed the basis for avoided costs, turned out to be much lower than previously forecast. With the California energy crisis there was renewed interest in renewable energy.

One crucial element of California's energy crisis resulted in the QFs, many of them small businesses, not being paid during the crisis years for the energy they generated. To help ease California's energy crisis, the QFs agreed to help stabilize short-run energy prices by locking in lower, fixed rates for five years, easing the utilities' cash flow and thereby helping the utilities to weather the energy crisis. That immediately reduced short-term energy prices paid to QFs by up to 50 percent. However, the legislature needed to take action to approve the plan and provide assurance that the QFs would be paid for past and future energy deliveries to Californians. California's 688 QF energy producers generate 30 percent of California's electrical energy requirements, enough to power 8 million homes.

III: Dusk—Year 2000

In the early 1990s, experts were estimating a need for 1,000 to 1,200 Mw of generation to be brought online annually, but the talk of deregulation discouraged utilities from investing in any new power plants. Officials anticipated being able to avoid rolling blackouts in 2001 and 2002 as some new natural gas fired power plants, which the CEC had recently licensed, were to come online. However, no new hydroelectric plants, nuclear plants, or other fossil fuel fired plants were being considered due to environmental concerns. Renewable sources of energy such as wind, solar, biomass, and geothermal plants contributed only 11 percent of California's electrical energy, and those sources were encouraged to increase in the immediate future. Governmental agencies had no authority to order the construction of new power plants.

In drafting AB 1890, considerable rhetoric had focused on the dread that the state's three major utilities would find ways to keep wielding monopoly power over the consumer. The restructuring was designed to forestall that possibility by forcing the utilities to sell their fossil-fuel power plants, which allowed new generators to enter into california's market.

With the passage of AB 1890 in 1996, construction of power plants in California slowed. Once deregulation became a factor, utilities stopped building power plants they believed they would eventually be required to divest themselves of. The utilities then had to rely on the power suppliers for their energy, and those suppliers then had to acquaint themselves with California's filing and environmental regulations for the siting, construction, and operations of power plants.

AB 1890 also required the regulated investor-owned utilities' rates for agricultural, residential, industrial, and large commercial customers to be frozen at the June 1996 levels and remain frozen until the utilities recovered their generation-related uneconomic costs through a competitive transition charge or until March 31, 2002, whichever came first. Starting January 1, 1998, rates for residential and small commercial customers were reduced by 10 percent and were to remain at that level until utilities recovered their generation-related uneconomic costs through the competitive transition charge or until March 31, 2002, whichever came first.

No one considered the effect on the market that manipulation from unregulated power providers could produce—their ability, as well as their incentive, to drive prices well above what could be expected in a truly competitive market. The bridled utilities were driven to the brink of bankruptcy, while the unregulated power producers chuckled all the way to the bank. Before the 1996 deregulation, utilities had been considered the foe, and the focus had been to constrain the foe. The rate freeze was originally designed to allow the utilities to recover past costs. However, the rate freeze, along with the accompanying soaring wholesale prices, turned the financial tables on the utilities in a disastrous way.

Utility workers expressed concern about the future of their jobs under deregulation. Tom Adams, representing the Coalition of Utility Employees, had admonished a legislative committee that the electrical energy surplus the state was enjoying could eventually evaporate and that "California [would] face high electricity prices and scarce supply, possibly brownouts."

The PUC had contributed to the problem by constraining the utilities from entering into long-term energy contracts that could have sheltered the utilities from wildly fluctuating prices. The PUC had been concerned that the utilities would be locked into high-priced contracts in a wholesale market where prices might fall below the high contract prices. As it turned out, the PUC had duped itself into believing everything should be done on the spot market. Deregulation stipulated that the utilities were to buy their energy from a new entity, the PX, which was basically a clearinghouse for electrical energy.

In response to AB 1890, PG&E overhauled its corporate structure, taking advantage of the opportunities it perceived in the national energy market. It established a parent company, PG&E Corporation, making PG&E (the utility) a subsidiary. Then PG&E Corporation proceeded to establish another subsidiary, the National Energy Group, to house its national business that was unregulated by California, a maneuver known as "ring fencing." In the subsequent few years, PG&E Corporation built a portfolio of businesses that included PG&E Transmission, which owned 612 miles of pipeline between Canada and the California-Oregon border; PG&E Generating, which developed and opened thirty power plants in ten states, mostly in the Northeast; and PG&E Trading, a market trader headquartered in Maryland, swapping billions of dollars worth of natural gas and electrical power. PG&E, the utility, after selling many of its fossil-fuel plants, as required by AB 1890, retained certain generating facilities, including sixty-eight hydroelectric plants, the Diablo Canyon Nuclear Plant, and two fossil-fuel plants.

Generating costs increased dramatically in the summer of 2000, resulting in PG&E paying much higher prices for energy than the rate freeze allowed. PG&E initiated a proposal to end the retail rate freeze and move the company out of the red and into the black. PG&E's proposal was to sell its most valuable assets, the system of hydroelectric dams located throughout the Northern California Sierra Nevada area, at an arbitrary asking price of $2.8 billion. The proposed buyer was to be another division of PG&E Corporation, PG&E's parent company. The deal would have the effect of simply transferring the asset from one division of PG&E Corporation to another; PG&E then could have claimed to complete a multibillion-dollar transaction, which would have canceled out all debts. The proposed transfer would have moved the hydroelectric dam's production out from under the purview of the PUC to a division exempt from state oversight. The U.S. Forest Service declared that PG&E's hydroelectric facilities were positioned among the most endangered habitat in the Sierra Nevada area, and the proposed transfer to unregulated ownership would have made the endangered area even more vulnerable.

SDG&E was able to pass its wholesale costs on to the consumer, raising the average household energy bill from about $50 to $120, thereby straining the budgets of businesses and schools. Surprisingly, a huge consumer conservation response was such that SDG&E's peak was the lowest since 1997, despite a steady 1.5 percent annual increase in electrical hookups. SDG&E's overall electrical energy use dropped by about 5 percent. SDG&E estimated conservation to be about 200 Mw per day. According to the ISO, the lower consumption helped to suppress statewide peaks.

As summer of 2000 faded, concerns over the supply of electrical energy began to surface. On August 16 the 2000 peak was 43,784 Mw and included 1,710 Mw of interruptible power that had been taken off line. The adjusted peak, including the interruptible power, could actually have been 45,494 Mw. For some time the state's hydroelectric supplies had been tapering off. California's power requirements were further aggravated by inadequate Path 15 thermal transmission capacity between Northern and Southern California. The result was an unavoidable Stage 2 emergency alert from the ISO. The ISO also invoked a federal power order that required suppliers in the West to sell power to California to meet their current demand. The ISO was having cash flow problems of its own, and the northwestern suppliers were reluctant to sell energy without assurances that they would receive payment, resulting in 10,000 to 11,000 Mw of unplanned deficiency. The DOE stepped in and directed the northwestern suppliers to sell to California. A temporary reprieve resulted from cooler than normal weather, contributing to a reduction of peak power requirements. While the peak dropped, the vital out-of-state energy supply sources dropped even faster. For years in the summer, California had comfortably relied on its neighbors in the Northwest and Southwest to supply power to the state as needed. That summer, California had been relying on 8,000 Mw of power to be imported from those neighboring states, but available out-of-state supplies dropped to between 4,000 and 5,000 Mw. Unfortunately, the state's Northwest neighbors were experiencing a drought. Ultimately, as well, the Southwest's growth caught up with their states' energy requirements, and California's lost that energy source.

Californians sat up and took notice on June 14, when the Bay Area was struck with rolling blackouts. There was considerable concern that if the peak continued to climb and if supplies from out-of-state suppliers diminished, Californians could expect additional periods of blackouts.

After a fivefold price increase in electrical wholesale prices, state regulators held FERC responsible for its refusal to cap wholesale prices. A PUC commissioner said the FERC had put the PUC in the position where any decision would be a bad one. PUC president Loretta Lynch vowed to keep the rates "reasonable," regardless of the utilities' assertions that they were restricted from passing on increased wholesale prices to the consumer.

It took repeated threats of rolling blackouts and spiraling wholesale prices before the PUC began expedited efforts to review utilities' requests for emergency rate hikes. PG&E and SCE informed the PUC that record-breaking wholesale electrical energy costs had pushed them to the edge of bankruptcy. They were out of credit and very close to being out of cash, and the utilities made it clear that the chaos was everyone's problem, stressing that the overriding public interest was to keep the lights on.

Any action the PUC took was unlikely to have had any impact on Standard & Poor's action in lowering PG&E's corporate credit rating to speculative or junk bond status. The utilities appeared to be headed for insolvency in a matter of weeks. SCE threatened to declare Chapter 7 bankruptcy if the PUC did not act decisively on rate hikes.

Senate President pro Tem John Burton (D-San Francisco) said the legislators would most likely expect something in exchange for a rate hike. One idea he suggested was the appropriation of the transmission lines. Other ideas included major changes in the structure of California's electrical market or a return to regulation. Consumer groups wanted any solution that would ensure California became energy sufficient.

In a PUC meeting, consumer representatives suggested that bankruptcy might do less damage in the long run than any hasty action in the present. The Utility Reform Network said they were in attendance because of an "alleged" emergency. Consumer advocate James Weil observed that it was disturbing that SCE had transferred $3 billion to its parent company in recent years and that PG&E had transferred considerably more. He commented that if the two utilities went bankrupt, it was "because their holding companies [were] letting them."

PG&E had a net worth of about $5.4 billion and needed $4 billion to pay for power energy costs that weren't covered by frozen rates or offset by the power plants it owned. The utility publicized a need for $7 billion, but at least 40 percent of that was for power from its own generating facilities.

In September of 2000, energy monopolized the political scene as heating oil supplies grew lean and natural gas prices soared. The Department of Energy released 30 million barrels of crude oil from the nation's Strategic Petroleum Reserve in an effort to relieve the heating oil crunch. Unfortunately, the country had limited refining capacity, so dipping into the reserve may not have helped the immediate shortage. The Arctic National Wildlife Reserve, located next to the Prudoe Bay area, was regarded as the most unexplored site for oil on the continent, and the Bush administration believed it should be explored. The rise in natural gas prices did not result in much support for ANWR exploration. Californians had been passionately opposed to drilling off their coast, and there wasn't much fervor to drill in the Gulf of Mexico. In October California's energy demand dropped for the first time in years, and no one really knew why. The deflated power peaks would have been welcomed in the following year.

With soaring wholesale prices, the PUC agreed on December 21 that consumers were going to have to face the reality of higher electrical energy rates, but that did not answer the questions as to how large the rate increases would be and when they would be initiated. The PUC said it would make a decision on rate hikes in a scheduled January 4, 2001 meeting, but the utilities had asked for immediate action. The suggested utility rate hikes ranged from 10 percent to 30 percent. PG&E had asked for a residential rate hike of 17.5 percent that could be increased up to three times annually over a five-year period. SCE increased its initial rate hike request from 10 percent to 30 percent. The PUC ordered an independent audit of all the utilities' books.

California's energy debacle turned more desperate on December 29. PG&E announced that gas consumers were going to pay two times what they had paid the previous year. Average residential bills were anticipated to rise from $50 to $125 for January 2001. PG&E informed the PUC that an initial round of its proposed electrical rate increases would cover only a fraction of the billions of dollars it estimated it needed over the long term. A PUC member opined that if PG&E's proposed rate plans were approved, electrical bills would triple in two to five years. The Utility Reform Network (TURN) called the gas price increases outlandish and said they expected January's bills to be higher than December's bills, which were already higher than the previous year. PG&E was allowed to charge a natural gas delivery fee that averaged about 3.6¢ per therm (100,000 BTU), and on top of those delivery charges it was allowed to pass along to the consumer the wholesale price, which was about 3.1¢ per therm the previous January. In December the price had risen to 7.5¢ per therm, and in the following month, it was expected to approach $1.42 per therm. PG&E faulted the PUC for allowing the high prices by discouraging the utility from building more gas storage space or investing in pipeline capacity rights.

Governor Davis asked President Clinton for emergency action to keep natural gas flowing into the state. In a letter to Clinton, Davis said, "This eminent gas shortage endangers the supply of natural gas for both residential and business customers and also for service to electric generating plants in Northern and Central California." The linchpin of the issue was whether or not enough energy could be secured for the long term at the right price.

Some other factors came into play. Governor Davis sought an extension of the DOE's order that compelled power producers to sell to California during the crisis. DOE's Bill Richardson did extend the order that would have expired at midnight, February 6, 2000. However, a lower demand in December was attributed to businesses' lower energy demand during the holidays, as well as the cooler weather.

The PUC was looking for creative ideas to help unravel the rate problem, and plenty of ideas were offered in December of 2000. Governor Davis started reaching out to consumer groups, looking for ideas to employ toward a solution. Jason Zeller of the PUC's Office of Ratepayers Advocates argued that the utilities, having the legal right to condemn property, should use their power of eminent domain to take over the power plants that were charging extremely high wholesale rates. While that may have been possible, the price of such a takeover would have been extremely expensive and not very practical.

Assembly Speaker pro Tem Fred Keeley (D-Santa Cruz) suggested that PG&E could meet its needed cash by selling its hydroelectric system to the State. A union coalition suggested a rate increase as a way to prevent power outages. Businesses and consumer groups urged the PUC to consider loans or other interim changes that could be undone later, when more information was available. But the PUC needed to focus on what was and what was not within its purview.

The PUC and the legislature were looking at price controls, anticipating a decision in January 2001, and were in the business of reviewing the energy the PX was buying on the spot market and evaluating whether that cost should be passed on to the consumer. They opined that the deregulated market appeared unsuccessful.

Several PUC commissioners indicated they favored emergency action to permit coverage of future wholesale costs, but they would not permit PG&E to back-bill consumers for the costs that occurred during the current rate freeze. In December, James Asselstine, a research director for

Lehman Brothers in New York, said the PUC really had only one chance to address the issue. Unless the commission voted to allow a rate increase to guarantee that PG&E could collect $4 billion from its ratepayers, the utility would very probably go bankrupt. He warned that if PG&E and SCE went bankrupt, the PUC would risk loss of its jurisdiction over power plants that the utilities still owned because of a possible court order for their sale. Asselstine warned the PUC that if PG&E was allowed to collect only future wholesale costs, financiers might be unwilling to lend the utilities enough money to keep buying power.

SCE sued the FERC, charging it with failure to control wholesale prices and noting that it had pushed SCE to the brink of financial ruin. Davis took his complaints with FERC to an interview on CNN's *Inside Politics* program, telling the country the Clinton-appointed agency had been very unresponsive to the state's energy crisis. Davis also said he was pursuing a transition to re-regulation that would include a price cap, a modest rate increase, and stepped up conservation.

IV. Dead of Night—Year 2001

January

As the state entered the year 2001, still in the deep throes of an energy crisis, January was filled with significant activity by everyone related to the energy business, and so the search for electrical energy continued.

On January 4, the PUC unanimously ordered a second round of hearings beginning January 8 to seek long-term solutions to the state's energy crisis. In addition, the PUC decided it would look at the possibility of legally condemning the power plants that were charging staggering wholesale rates. The use of State-authorized bonds as a way to reduce utility debt was also considered. Consumer groups pleaded with the PUC not to increase rates and to simply allow the utilities to go bankrupt.

The PUC approved an emergency electrical energy rate increase, limited to ninety days. Residential rates were increased by 9 percent, small business rates by 7 percent, and large company rates by 12 percent. The increases were well below what the utilities had requested—they had asked for three times the amount granted. The rate hike took effect immediately, just three years and three days after the January 1, 1998, 10 percent rate cut. The temporary rate increase was projected to produce an additional $200 million each for PG&E and SCE over the next three months.

Financial analysts said the rate increase was far below what the utilities had been requesting, and the rate action sent utility stocks plummeting. The utilities needed more than $10 billion to cover what they had paid for wholesale power since May 2000. PG&E stated it needed $4 billion to fill the gap between what it had collected in rates and what it had paid for wholesale power. It called the rate increase a first step but made it known it would now pursue other avenues to solve its cash crisis. The companies considered issuing bonds, which would be paid for by higher electric rates for a decade or more. PG&E's and SCE's bond ratings were lowered to junk bond status. Both PG&E and SCE sued the PUC, asking the federal courts to allow to them charge higher rates.

The rate increase hit Northern Californians particularly hard—they had already seen their natural gas bills increase without apparent limits. Residents' January utility bills that were as much as 60 percent higher than their December bills; the average residential bill rose from $50 to $125. As a result, the consumers felt they were being "ripped off." Farmers also got a taste of doubled natural gas prices when they paid the bills for their fertilizers and pesticides. Processors had their turn, paying more for the natural gas they normally used in their processing plants. The problem was compounded by a softening farm economy, with low commodity prices preventing the farmers and producers from passing on their higher costs.

An energy plan designed by the Regional Council of Rural Counties proposed reopening thirteen existing biomass cogeneration facilities that could produce 150 Mw. California at the time had twenty-nine cogeneration plants burning wood and tree wastes, producing 540 Mw. A lack of materials in the 1990s had contributed to the closure of several sawmills and their associated cogeneration facilities. Reopening the thirteen plants required a long-term commitment for a reliable supply of biomass, and that posed a problem. Most of the wood-fueled cogeneration plants come from the U.S. Forest Service, which owns 60 percent of California's timberland. Additionally, removing these fuels would provide fire protection. Cogeneration proponents were fearful the Forest Service would reduce the amount of available biomass. A new plan for managing national forests was designed to create a predictable supply of timber not subject to the dramatic ups and downs as experienced in the past. The new plan, amended under the framework for Sierra Nevada forests, was legally and scientifically defensible. Environmentalists generally endorsed the cogeneration facilities, as long as they complied with the Clean Air Act. The Rural Council proposed building sixty new cogeneration plants at sixty lumber mills that had closed during the last decade. Together they could provide up to 8,000 Mw. However, the cogeneration plan also depended on a commitment from the utility companies to buy the power at reasonable rates. Most cogeneration plants could operate at a profit if they sold their power for 5 – 8¢ per kwh. Prices on the spot market then ranged from 40 – 50¢ per kwh. Several plants were idled after PG&E Company bought existing contracts when their price dropped to less than 4¢ per kwh.

Two energy projects in Kern County were proposed—wind turbines in Jawbone Canyon and a $350 million plant south of California City. FPL Energy applied for permits to build 182 wind turbines on 1,200 acres in Jawbone Canyon, a project that would generate 270 Mw and be completed by the end of 2001.

SMUD observed that the Sacramento region badly needed either new power plants and/or transmission line improvements to meet its electrical energy requirements. SMUD predicted the region's electric grid could be stressed to near-collapse within a few years. SMUD hoped to put a new natural gas burning plant beside its closed Rancho Seco Nuclear Power Plant. SMUD sought permission from the CEC to build up to 1,000 Mw of power plant capacity. SMUD estimated the licensing process would cost about $2 million for the air permits, air pollution credits, and other requirements. A natural gas pipeline right-of-way might add $6 to $12 million to the cost. The SMUD peak usage at the time was 2,759 Mw.

David Holberg, civil engineer and writer for the Ayn Rand Institute said, "Green activists have worked for decades to stop the construction of major power plants in California, and they have succeeded. The recent blackouts are evidence that their goal is to protect fish, trees, and water at the expense of man."

The ISO said that all day on January 9th reserves were tight because 10,000 Mw had not been available for service due to planned and forced outages. Diablo Canyon Power Plant was down because the saltwater intakes were plugged. This plant accounted for one-third of PG&E's generating capacity. The ISO reported it was having problems importing power to keep California supplied with electricity, and as a result, the DWR eventually had to buy power from energy traders who had balked at selling to the ISO.

A problem with Path 15 developed in the two transmission lines from Los Banos to Midway. The line's limited thermal capacity was restricting the transmission of power from Southern to Northern California. To many it was emblematic of California's obsolete infrastructure, while others saw it as only a minor annoyance. Obviously, California's biggest shortcoming was the lack of power generation. Twice PG&E investigated adding another line, but the study couldn't justify the costs. The ISO and PG&E worked an alternative cost-benefit study they hoped would produce more positive results. Various power companies had considered building 3,000 Mw of power plants south of Path 15 but were deterred without an upgrade of Path 15, which was required to transmit the additional power.

Davis was prepared to take emergency action, including the takeover of power plants. Wholesale suppliers were still willing to sell to PG&E and SCE—if the legislature could come up with a viable solution to the crisis. AB 1x was a hastily drafted bill to assure impatient power generators and consumers that the DWR would be authorized to make long-term contracts for the purchase and delivery of electricity directly to the consumer. The bill would give the DWR authority to buy power at 5.5¢ per kwh. Under the bill, the DWR was given the authority to borrow money—about $400 million—from the general fund for startup costs. Assemblyman Keith Richman (R-Sun Valley) wanted the bill to be broader in scope and include other concepts, such as accelerating the process by which power plants were approved and built. Reliant Energy observed they were not seeing many concrete solutions.

Some legislators were concerned that with the DWR buying power, the State could be committed to hundreds of millions of dollars in energy costs that the State would never see again. They demanded a limit on the amount that could be spent, and they wanted assurances that long-term strategies, such as relaxing environmental review to speed up plant siting, would be pursued. There was concern that the purchase of power could potentially deplete any financial surplus that the State had achieved. Several sources said the State was paying as much as $800 per mwh, about fourteen times the price Davis believed Californians should be paying. The ISO estimated it was paying $10 million to $80 million daily for last-minute power purchases, and the DWR was spending $6 million each day. The pledge to have the DWR buy power pretty much calmed wholesalers from cutting off sales or filing bankruptcy petitions against the utilities.

Governor Davis learned on January 10 that the utilities were paying a credit penalty on electricity because of their dubious finances. Davis declined to make any commitments regarding the use of State-backed bonds to pay off the utilities' debt. Davis and the legislature said the State could solve the generators' problem of reluctance to sell to the utilities by one of two means—either the State could guarantee payment on long-term contracts or the State could buy power itself and then resell it to the utilities.

On October 23, 2000, Duke had submitted plans for a 1,200 Mw power plant "modernization" of its existing 1,030 Mw facility in Morro Bay. On November 29, 2000, the CEC voted the application as "data inadequate." On January 11 the CEC began a formal review of Duke's proposed Morro Bay power plant project. Steve Larsen, CEC executive director, eventually recommended acceptance of the plan as "data adequate." The project included the construction of 1,200 Mw of new combined-cycle units and demolition of the existing plant and stacks.

On January 11 PG&E was confronted with a new emergency—it soon might be unable to purchase natural gas for its customers due to lack of cash. With PG&E's deteriorating credit, sellers were reluctant to sell the utility any more gas. PG&E advised that their gas supply would be exhausted in the second week of February. PG&E asked Davis to use his emergency powers for help, such as buying gas or providing credit. Gas sellers had become nervous about their payments getting involved in a possible bankruptcy. On Tuesday, J. Aron & Company, a natural gas company owned by the trading firm Goldman Sachs, informed PG&E that they wouldn't sell them any more gas unless the utility paid cash. On Wednesday, PG&E met with suppliers to assure them that they could pay for gas supplies, since they were now able to pass gas costs on to consumers. PG&E also announced it would skip its regular fourth-quarter dividend in order to conserve its cash position.

January 12, state officials revealed a plan for the State to buy wholesale electricity and resell it to the utilities. The plan would include provisions for retirement of the nearly $12 billion of utilities' debt. In one scenario, the debt would be repaid over time by indefinitely extending the 10 percent, 90-day rate increase, which was approved the previous week by the PUC. Davis decided the plan was to be structured so the State was not financially at risk for the buying and selling of power. The State would buy power directly from the generators under long-term, high volume contracts. The State would then resell the power to the utilities at cost, plus a minor administrative fee. The DWR had already spent at least $25 million for power on the spot market. It was suggested that the generators should be allowed to relax air quality rules in exchange for a favorable power-purchasing deal, but the administration was not supportive. There was very little support by the legislature for a "securitization" plan under which the State would sell bonds that would retire the utilities' debt and be repaid by ratepayers' revenue. There was more support for a debt reduction plan that would maintain the 10 percent rate increase as long as the State received something in return from the utilities, such as generation facilities, transmission lines, or company stock. It was the difference between a bailout and a business deal. The Foundation for Taxpayer and Consumer Rights (FTCR) said the utilities should be required to pay interest.

The bond plan was one of many proposals to be debated in the following weeks.[12] Rod Wright (D-Los Angeles), chairman of the Assembly Utilities and Commerce Committee, believed the bond debate had a fundamental question yet to be resolved—was the debt the responsibility of the ratepayers or the shareholders? It was not clear who would be bailed out. An analysis released by the PUC's Office of Ratepayer Advocates indicated the bond issue could benefit the consumers and the utilities in two ways: (1) rate reduction bonds could be issued at lower costs than other bonds because they wouldn't be tied to the utilities' dwindling credit ratings; and (2) these could be taxed over many years, deferring what would otherwise be a major tax bite for the utilities.

Consumer groups said they would contest any plan to retire the debt if it involved the ratepayers' money for a public bailout of investor-owned companies that had made bad decisions in California's deregulated market. Senator Burton said he was willing to support the power of eminent domain to put the State in the power-generating business. The stumbling block became the question of retiring the utility debt. Davis suggested the utilities would have to relinquish transmission lines or other assets under any debt repayment deal.

On Saturday, January 13, Governor Davis and state legislature leaders participated in a marathon negotiating session of about thirty people. They emerged saying they would recommend the State step in and permit the purchasing of as much power as California needed to allow the utili-

---

[12] Senate President Pro Tem John Burton called the utility support deal "… no deal I'm a part of … if you assume the ratepayers are going to pay for this bailout, then do you charge them $20 a day for a year right now, or do you charge them $10 a day for two years or $2 a year for 10 years, or whatever."

ties to regain control of their financial footing. The meeting concluded without any agreement on any terms of long-term contracts with power generators or a solution to the long-term debts. Davis and the lawmakers were seeking to fix the situation through a plan where the State would buy power over the long term and resell it to the utilities. That would eliminate the premium prices that utilities were forced to pay due to their declining credit rating. The far more controversial part of the plan, which was still under discussion, would be the help needed to repay the utilities' billions of dollars of debt by extending indefinitely the 10 percent rate increase the PUC had granted for ninety days on January 4.

Key lawmakers acknowledged they couldn't guarantee that the State would be repaid in full as a result of AB 1x. Senator Burton said that although it was not the best solution, it was the only solution to the State's short-term problem. Problems were compounded by the reluctance of Canada and others to sell power to the State. U.S. Senator Dianne Feinstein (D-CA) warned that the crisis could ripple through the entire global economy. President Bush warned that California was pretty much on its own in resolving their crisis, saying, "California is going to have to address and correct the law that has caused some of this to happen."

On January 17, the ISO declared a Stage 3 emergency that affected about 500,000 residents and would continue until 9 p.m. the same evening. The ISO warned that they would give only ten-minute notices of outages. The fact that California's inadequate energy supply was a serious problem soon became very apparent.

On January 17, SCE and PG&E told the SEC that they had stopped paying their bills. They still had cash on hand and wanted to reserve it for their daily operations. Gary Ackerman of the Western Power Trading Forum said nothing sinister was going on—that it was just the way a business market works, that if buyers pay the market price, they'll get all the power they need.

PG&E had already borrowed $5 billion in order to keep supplying electrical energy. PG&E had proposed the idea of a bond issue to be paid off with increased costs to ratepayers for years to come. The Office of Ratepayers Advocates alternatively suggested that PG&E pay off the obligations with money it collected from the increased rates.

The PUC voted two to one to order the two California utilities to keep power flowing to their customers despite the companies' financial hemorrhaging. The utilities were considering providing no more power than they could generate. The PUC cited "the obligation to serve" law—you must continue to serve your customers as long as you have the ability. Earlier in the month, PG&E had filed a plan with the PUC that envisioned the utility being forced to cut off power to its customers for up to twelve hours a day.

A poll in the *Los Angeles Times* indicated that more than 54 percent of Californians didn't believe there was a power problem. Polls showed the majority of Californians suspected energy shortages were being manipulated to drive up electrical rates.

Then there occurred a cascade of other problems. A Wall Street credit rating agency put California on its Credit Watch list shortly after Davis signed a bill authorizing the DWR to spend $400 million on emergency power purchases. The Credit Watch was in essence a warning that the State's credit rating might be downgraded, making borrowing cost millions more. California's gasoline suppliers began to feel the electrical pinch. Kinder Morgan Energy, a major gasoline pipeline operator, had been shut down for up to hours a day for several days. This forced gasoline to be transported by truck, a very inefficient method resulting in increased prices. SCE edged closer to bankruptcy, defaulting on an additional $32 million worth of bond debt. It also served notice it didn't expect to make payments on $223 million of bond debt due at the end of the month. SCE advised it would run out of cash by February 2 if it didn't start skipping debt payments. The PX started to shut down. Nobody was using the market, and nobody would trade with the utilities, causing the PX to become nonfunctional. SCE sued the PX to prevent it from liquidating its power futures contracts, which had been obtained at attractive rates. The PX proposed selling the contracts because SCE had defaulted on a $215 million debt to the PX.

Environmental concerns did not help the search for new sources of natural gas or the need for additional storage capacity. A spokesperson for the PUC said, "We want to see more gas in California. If it means putting in more pipe, that might call for environmental review, and we would need to look at impacts for that." Environmental concerns dictated that all new power plants burn natural gas—the only fuel that could meet California air quality standards. The ARB was adamant about producing power without adding any pollution, and natural gas had become the fuel of choice. Power generators were required to purchase air credits to lower pollution.

January 18's random bout of blackouts began to elicit reactions that were less than gracious. Concerns developed that businesses might think California was a risky place to build as confidence in the state's electrical supply waned. Power generators took a disapproving view of any State plan to go into the business of buying electricity.

On January 19, Standard & Poor's listed California on its Credit Watch list because of the "uncertainties" surrounding the State's ability to fashion a long-term solution to the energy crisis and the financial effects that would likely follow if the State's credit was eventually downgraded and the cost of borrowing money increased. State Treasurer Phil Angelides said, "It's just another dose of reality." The Western Power Trading Forum said it appeared that the DWR would deplete its fund in five to eight days. Once more there was considerable talk of the utilities selling their assets to pay off the debt.

On January 21, California experienced its first blackouts, which consumer advocates considered to be finance driven. Consumer advocates said power plant breakdowns, jammed transmission lines, and low rainfall would have been irrelevant if enough money was on the table—they believed California did not have a power shortage. In the summer months, California typically experienced a peak in excess of 45,000 Mw, and in the winter the experience ran 30,000 to 35,000 Mw. For example, as a result of the January 4 rate increase, PG&E received about 12¢ per kwh from its residential customers, and of that, 6.5¢ went for buying electricity. At the same time, the wholesale cost exceeded $1 per kwh. There was no question that the utilities went several billion dollars into debt.

On January 22, a special auction began for sealed bids from power producers for long-term contracts that would last from six months to ten years. Davis and key lawmakers continued taking in sealed bids behind closed doors until noon, January 23. It was questionable whether the wholesalers would accept Davis's proposed price of 5.5¢ per kwh. The guess was most wholesalers would bid 7.5¢ per kwh for contracts of about five years or so, but would probably insist on contracts as long as ten years, the maximum allowed time. One item of considerable apprehension in the bidding documents was the sentence, "Neither the full faith and credit nor taxing power of the State of California were to be pledged for any payment under this agreement resulting from this solicitation."

Results of the twenty-seven-hour bidding process conducted by the DWR during January 22 and 23 produced bids from generators with a "weighted average" of 6.9¢ per kwh. Davis had set 5.5¢ as the maximum price needed to avoid rate increases. Lawmakers observed that the bids gave them a better basis on which to craft a bill that would authorize the State's entrance into long-term contracts. Some wholesale generators said the inflexibility of the bidding rules would effectively increase the terms of their bids. The bidding rules included the State's insistence on firm prices without allowance for changes in natural gas prices. Given the uncertainties of future gas prices, wholesalers were motivated to raise the bid price as a way of reducing their risk. The State's bid system didn't allow for any creative bid alternatives that might have resulted in lower prices. Reliant Energy met the bidding rules with a compliant bid for a limited amount of power, but it included an alternative bid for far more power, with the stipulation that the State purchases the natural gas.

The average bid for long-term contracts of 6.9¢ per kwh was nearly 40 percent higher than the price the utilities could have locked in 2000 had not the PUC rejected it. The 6.9¢ price was artificially low because it only averaged the prices of non-peak power. It is worthy to note that in 2000 the utilities had the opportunity to lock long-term contracts at about 5¢ per kwh, but the PUC had rejected those contracts. The price at the time was running at 30¢ or more on the spot market during peak times.

Davis administration officials predicted that an agreement on a long-term solution would be accomplished. Auditors from the PUC, state senate, and Credit Suisse met to seek a mutual agreement regarding the utilities' current cash position, their debt, and potential future costs. In Washington, members of a California delegation introduced legislation that asked for price caps, but the bill did not receive any prompt action. State Controller Kathleen Connell said she would back legislation SB 13x to place liens on utilities' assets to ensure that the State would be repaid the $400 million the lawmakers had authorized DWR to spend on power purchases through the end of the month.

Under PG&E contracts, interruptible customers could be cut off for only one hundred hours a year, which time had already been consumed by January 22. So even with possible blackouts looming, the ISO lost its ability to tell about 170 big interruptible customers that they must either lose electricity or pay big penalties.

There was a flurry of activity in Washington on January 22. The new Bush administration sent a strong message that it was not open to capping wholesale power prices. PG&E continued to lobby the administration to extend a federal order that forced natural gas firms to continue supplying natural gas. The ISO and others pleaded for federal extension of emergency electricity sales.

On January 23, the DOE extended emergency orders to keep power flowing to the state until February 7 but said it would be the last. DOE's Abraham said, "Our action today is designed to give the governor, the California legislature and other relevant parties the time to take the necessary action; I urge them to act immediately . . . While we do not intend to write legislation for the state of California, the real solution must address the need for more electrical power generation in California, flawed state market rules, restoration of the financial health of California utilities, and encouragement of conservation."

SCE had warned of the "clear and present danger" of widespread blackouts and urged a federal judge to order a 30 percent rate increase throughout SCE's service area. In a filing, SCE accused state officials of refusing to recognize the reality about the company's wholesale costs, saying that the PUC's "timid action has accelerated the financial crisis." SCE claimed that it "must be permitted to recover its wholesale costs."

SCE was extraordinarily disappointed that its sought-after rate hikes were not granted. The SCE increases would have started at 30 percent and grown as high as 76 percent in two years. SCE raised the possibility that there would be blackouts in Southern California if it were unable to borrow adequate money to purchase wholesale electricity. The PUC was skeptical, as the distrust of utilities was so widespread; one commissioner opined that it would not be surprising if the utilities allowed some blackouts. This speculation was common among consumer advocacy groups.

SCE requested that the PUC relieve it of its legal duty to supply power to everyone. That request raised many eyebrows; SCE was able to meet 80 percent of its own power requirements. Additionally, to economize, SCE said it planned to trim its work force by 1,450 jobs over the upcoming several months. The previous month, SCE had eliminated 400 jobs. The total loss of 1,850 jobs represented 14 percent of its 13,540 work force.

In a filing, SCE served notice that it had temporarily suspended payment on $596 million in debts that were due—$230 million to bondholders, $151 million to some power wholesalers, and $215 million to the PX. Moody quickly downgraded SCE to junk bond status.

SCE was allowed two additional days to pay a $215 million debt to the PX, and SCE decided not to make the payment, scaring off suppliers. The

SCE move prompted the PX to consider selling off future supplies that SCE had contracted for at attractive prices. PG&E announced that its inability to buy natural gas supplies had reached a "crisis level." Gas companies that represented more than 40 percent of PG&E's supply threatened to cut off deliveries because PG&E was unable to put cash up front.

There had been talk of requiring the utilities to relinquish some of their assets for a bailout. A director of Edison International, the parent company of SCE, opined, "The utility shouldn't have to give the State any financial assets in order to extinguish a debt that was forced upon SCE by state officials in the first place."

President Clinton, in his last day in office, declared a "natural gas supply emergency" in Northern and Central California, enabling the DOE to order gas companies to continue supplying product. Two suppliers, however, had already halted shipment to PG&E, demanding cash up front. In fact, two-thirds of PG&E suppliers were considering curtailing deliveries.

Several suppliers that accounted for 36 percent of California's daily natural gas supply had previously announced their intention to stop delivery as of January 23. Other suppliers that accounted for another 30 percent of the daily natural gas supply issued the same ultimatum—until the DOE ordered the continued shipping of gas. California does have its own limited gas resources; however, the majority of its gas comes from the Southwest and Canada. The PUC considered an order against the two utilities to require continued service, noting that the failure of reliable service could cause "serious jeopardy and irreparable harm." PG&E was very much dismayed at the statement, saying, "That implies we would abandon service to them." They went on to say, "At a time when we need real solutions to the state's energy crisis, the Commission had wasted precious time chasing after a problem that did not exist. They continue to ignore the fundamental problem that we could no longer afford to buy power for our customers."

The PUC was looking into the possibility of price manipulation in the natural gas market. In April of 2000 the PUC filed a complaint with the FERC, alleging that El Paso Natural Gas Company was withholding gas at the California border.

PG&E denounced wholesalers for charging, "whatever the market would bear" and began pricing energy from its own plants at five times the cost of production, for accounting purposes. PG&E indicated the new price could not be passed on to consumers while rates were frozen. In a filing to the PUC, PG&E elected to price its nuclear and hydroelectric power plants at 15¢ per kwh—15¢ was selected because that matched the proposed price cap selected by the FERC. PG&E was now unable to use its old formula that had been tied to the PX prices because now the DWR was buying power, and PG&E no longer traded on the PX. The PX's price had varied higher and lower than 15¢. Production of electricity generally cost about 1¢ per kwh, and previous PUC analyses had pegged PG&E's nuclear costs at 3¢ per kwh. More than 11,000 Mw of hydroelectric, nuclear, and coal-burning facilities were still owned by PG&E and SCE. Regulators and others suggested that some lower costs could be realized by taking the plants still owned by the utilities out of the expensive wholesale market. PG&E did not take the position that it should go back to the old way of setting power rates—at cost plus a regulated profit margin—until the PUC lifted a State-ordered rate freeze. PG&E was unquestionably ready to pledge its generation at cost-based prices as part of a larger solution, but until other fixes were in place, PG&E believed its power should be paid for at market prices. PG&E was ready to go for market prices, but only if the PUC approved PG&E's rate stabilization plan, which would have required residential rates to be increased up to 40 percent in the first year, coupled with the allowance of twelve more rate hikes in the following four years. PG&E lobbied to preserve the old agreement that half of any above-costs on Diablo Canyon should be passed on to its shareholders.

Davis and the utilities petitioned the FERC to put a cap on the price of wholesale power but were rejected. Meanwhile, a coalition of consumer groups endorsed Davis's proposal to create a public power authority to build more generating facilities.

A meeting of the Western Governor's Association in Portland on January 25 and 26 quickly concluded there wasn't enough power generation in the West to meet the region's growing population and economies, and fault for that was not only on California—it was a regional problem. The region was not building enough generation to keep up with the growth. California's electricity crisis had been a jolt to the western states. Increased power costs had contributed to the bankruptcy of an aluminum plant in Vancouver, Washington by forcing its utility bills up by 50 percent. In Tacoma, Vanalco, a company in business since 1940, was forced to shut down, with a loss of 650 jobs. Scarce power jeopardized 2,350 mining jobs at Phelps Dodge in Arizona and New Mexico. Miller Brewing Company in Southern California laid off 160 workers and moved its production out of the state.

California remained dependent upon supplies from outside the state because of its failure to build any new generation facilities over the preceding twenty years. California would eventually become less reliant on out-of-state sources, as nine new power plants in California had been approved for construction by the CEC—five power plants were already under construction with a capacity of 3,988 Mw. The CEC expected 2,368 Mw would be available by the end of the year, the balance being anticipated by 2003. However, the additional generation was not expected to impact energy prices because all of these new plants were fueled by natural gas, whose prices had been rising as much as electrical prices.

In an effort to address the electrical energy shortage of the past year, suppliers from San Diego to Redding ran their generating plants more intensely and frequently, firing up their polluting, standby turbine generators to keep supplying electrical energy. To meet the energy squeeze, power managers relied on a network of several hundred fossil-fuel generating plants in periods of heavy demand, plants which were aging and ranked among the largest polluters in the state.

A review was made of twenty-one power plants that were proposed or were under construction in California; it reported their progress was hampered for a variety of reasons. Environmental activists had slowed or stopped some power plants, as well as had labor unions, corporate neighbors, and others with abundant resources at their disposal. Power generators complained the loudest about new plants, in an attempt to hold up the licensing of their competitors. Of the thirty-one plants proposed for licensing since 1997, competing companies had intervened in twelve proposals, slowing the process in at least four instances. Power producers had an interest in all of these cases. Producers also had to deal with constrained resources such as natural gas and transmission line capacity. Across the state, California's NIMBYs had also applied pressure. State lawmakers pressured regulators to bypass community concerns and speed up the licensing of new power plants. Wall Street editorial writer Susan Lee described California as "really rich," taking power from other states because California didn't want to build any unsightly power plants. Cisco officials, for example, didn't want a power plant's blight next to their proposed corporate campus, which they had been planning in the San Jose Coyote area.

The CEC tried to speed up new generation with its "one-stop-shop" methods for permits. The result—the CEC licensed nine big power plants totaling 6,278 Mw. Six of those plants were under construction, and firms were formally seeking licenses for twelve others. Three times in the last decade, the CEC had approved plants that were never built. To build a big gas fired power plant in California, the applicants had to go through at least twelve months of NOI and AFC public hearings, evidentiary hearings, and cross examinations. The results had to be reviewed in preliminary staff reports, and hundreds of public hearings had to be presented before the CEC could reach a decision. The process originally took about eighteen months. Other states licensed plants faster; a few of the states had severe smog problems or shortages of available land. CEC commissioners had sought to fast-track power plants by offering a six-month review for certain pre-approved sites. They also required state environmental agencies to speed up their comments.

Calpine, with six plants in the works, complained about the growing emergence of professional interveners who would go after all these power plants. Two groups were expected to raise more objections than others: Californians for Renewable Energy (CRE) and California Unions for Reliable Energy (CURE). Enron said power plant licensing was about twice as slow in California as it was in Texas.

According to the CEC, twenty-two energy companies had intervened in twelve competitors' proposals. Some became formal interveners simply to glean information. Others aggressively used their status to hire attorneys, file objections, and cross-examine competitors in evidentiary hearings. In Pittsburg, the Southern Company and Enron intervened in the Delta Plant proposed by Calpine; they were intractably worried it "would clog up available transmission lines" and deplete the availability of natural gas. The Calpine plant eventually was licensed, but the process took sixteen months.

In San Diego County, Duke, Sempra, and NRG raised concerns that Calpine's proposed Otay Mesa plant would take away natural gas and transmission capacity from their operations. In Sutter County, rice farmers raised concerns over Calpine's 500 Mw plant, ostensibly being afraid of crop-dusting planes becoming entangled in Calpine's transmission lines. Others questioned, would the plant be noisy, smelly, floodlit at night, and would it use up available water supplies? In an effort to placate those issues, Calpine agreed to limit its water pumping, dampen noise, and plant 680 redwoods to hide the $300 million power plant from its surroundings. Sixteen months passed before the plant was licensed. Then after construction had started, a San Francisco woman who owned Sutter land filed an appeal with the EPA. Another four months went by before construction was allowed to resume.

January events once again revealed how the state's energy crisis solutions remained both elusive and constantly changing. The Democratic assembly thought they had an answer, but it quickly ran aground in the senate and in the financial community. The Democrats proposed to seek long-term bids for power and resell it to users through the utility companies. The utilities were to dedicate to the State some of their assets, either hydroelectric plants or transmission lines. The utilities would then be allowed to recover their losses by extending the average 10 percent rate increase granted by the PUC.

Some lawmakers balked at taking over the utilities' hydroelectric generating facilities, while others said they didn't approve of any idea that put taxpayers or ratepayers at risk for the utility companies' debt. Senator Burton maintained any long-term solution produced by the senate must contain stringent protections for the ratepayers and taxpayers. Utility lawyers and financial analysts said any such plan could run afoul of existing financial instruments, under which utility assets had already been pledged as collateral. Burton introduced a bill to create a public power authority with the authority to sell revenue bonds to finance new power plants and offer low interest loans to consumers who purchased energy-efficient appliances. The bill would have required the bond-financed generating plants to sell power to California at cost.

State Treasurer Angelides proposed a plan that would allow the California Consumer Power and Conservation Authority to (1) issue up to $10 billion in State bonds to finance conservation programs through low interest loans, (2) purchase the transmission facilities from the investor owned utilities, and (3) construct new power plants. Initially, it would have cost the general fund $1.2 billion, which was to be theoretically repaid as loans to the financing authority. Angelides said the power generating plants would ensure California would never again be subjected to the wild fluctuations of the private energy market. Burton observed that the idea made a great deal of sense. Senator Tom McClintock (R-Simi Valley) indicated he would back the plan as long as it did not include re-regulation.

There was talk of sending the utilities into bankruptcy. However, a bankruptcy judge's control over much of California's energy supply would present uncertainties that policy makers were loath to deal with. A bankruptcy judge's duty is to the creditors, not to the ratepayers. Utility bankruptcy had the potential to damage the state's economy as well as exacerbate the U.S. economy's slowdown. In his State of the State address, Davis admonished the public that bankruptcy would mean millions of Californians would be subject to blackouts, businesses would close, jobs would be lost, and investment would flee the state. While the California economy could potentially suffer a devastating blow, Wall Street analysts criticized Davis for failing to outline a detailed plan to deal with the utilities' cash crisis.

On January 31, the state senate passed the measure to allow the State to enter into long-term negotiations to purchase electricity. The legislation would (1) authorize the DWR to enter into power contracts over a two-year period, until January 2, 2003; (2) allow the DWR to issue revenue bonds, not exceeding $10 billion, which were to be repaid through utility rates; (3) establish a start-up fund of $500 million borrowed from the State general fund to be repaid at the earliest possible time; (4) allow the State to buy power without competitive bidding; and (5) prohibit rate increases for residential customers who did not exceed the threshold amount. Senator Bill Morrow (R-Oceanside) commented, "I don't relish being the messenger of doom. This bill does nothing more than extend the exact same recipe that brought the investor-owned utilities to the brink of bankruptcy."

February

In February, the utilities were way ahead of the state officials in seeing potential wholesale rate increases on the horizon. SCE had seen the problem as early as March 1999, when no one else seemed to have a clue. SCE asked the PUC to allow them to enter into long-term energy contracts that they had negotiated directly with suppliers, but the PUC refused the request. To exacerbate the problem, SCE had asked the PUC seven times to allow it to enter into contracts similar to the governor's currently negotiated contracts. Each time, the PUC denied SCE the contracts. And Davis's new solution of long-term contracts came much too late to save the utilities from debt. The PUC had feared that by allowing the utilities to enter into contracts directly with the suppliers, the utilities would be able to corner the market and stifle competition. The PUC was not alone in that belief—every conceivable group agreed with them, including California's largest energy users: the California Manufacturing and Technological Association, the Farm Bureau, the governor's department of General Services, and representatives from the TURN. The PUC did not recognize that the utilities did have a significant incentive to negotiate the lowest possible energy prices—self-preservation. Now all those low prices that had been available were lost.

In exercising his emergency powers, Davis seized low-cost energy contracts from the PX without knowing the price. On February 2 he seized a group of contracts between SCE and SCE's power suppliers, and on February 5 he seized contracts between PG&E and PG&E's various power providers that had been negotiated in 2000. Davis was subsequently billed $1 billion for the contracts that he had seized. All of these contracts were substantially below the spot market prices. Davis said the money for the purchases would initially come from the general fund and would be repaid by the sale of power to the consumers. The PX invoiced the State for its seized contracts: $651 million for the SCE contracts and $347.9 million for PG&E's. PG&E had contracted for as much as 500 Mw per day, with prices ranging from 6¢ to 13¢ per kwh. SCE had initially paid $215 million for their contracts. The PX said the invoices represented the "reassessed values of those contracts based on the current market conditions," and the PX expected to be reimbursed for the full invoiced amount. The courts had previously given the PX permission to auction the contracts before Davis stepped in, and a court hearing had been set for the PG&E auctions. Regardless of the PX invoice, it was not clear how much the State would eventually pay for the twelve-month contracts.

Governor Davis announced on February 5 that the State had reached agreements for long-term power from more than twenty suppliers. He said the agreements were cheap enough to fit within the utilities' "existing rate structure," ignoring that the structure had already allowed for a 20 percent rate increase over the next year. Davis said he had actually lined up forty agreements, or about 75 percent of the supplies needed for the next several years. He extolled the agreements as a breakthrough in his efforts to fix California's electricity problems. Some of the agreements were to last as long as twenty years. Consumer advocates and independent analysts believed he was overplaying his hand.

Governor Davis announced on February 6 that his energy negotiators had reached agreements on the State's first long-term energy contracts, but did not release any of the details. The long-term contracts provided for up to 5,000 Mw of electricity. The *Sacramento Bee* filed a public records request seeking the prices, vendors, and terms of all of the purchases the State had made starting from January 1, regardless of if they were the spot market or through long-term contracts.

On February 7, the DOE let the order to keep supplies flowing to California expire.

Davis announced on February 8 that he was prepared to use his emergency powers to permit expedited approval and construction of power plants. This authority had been granted to the governor by the State Emergency Act and allows the governor to declare a state of emergency and invoke a wide range of emergency powers when a natural or man-made disaster threatens the safety of Californians. He is allowed to suspend state laws and use State money to commandeer private property; the value of the property to be determined after it is seized. Commandeering private property differs from the use of the power of eminent domain. The act specifically permits a state of emergency to be declared during a "sudden and severe energy shortage." Davis made it known that he intended to use his emergency powers to boost electrical output of existing facilities, accelerate power plant construction, shorten the review process, and provide incentives in the form of tax credits to encourage renewable generation.

The western states held a meeting in Portland, Oregon on February 10 to seek a consensus on a plan to lobby Congress and the Bush administration for help during the energy crisis. Washington's Governor Gary Locke did recognize that it was not just California's problem. However, other governors prodded California to do more in the long-term, and in particular, streamline the permitting process. The governors realized that they all had to adjust their guidelines, at least temporarily, and that California should be the first to come to the party. The majority of the governors were seeking temporary control over wholesale prices. The cap they were seeking was the cost of power production plus a "reasonable" profit. Curt Herbert, chairman of the FERC, cautioned that temporary caps would act as a disincentive to both conservation efforts and the construction of new power plants. Additionally, the caps would discourage power buyers from leaving the volatile spot market in favor of entering into long-term contracts.

Confusion in the state was compounded by many who believed Davis failed to assume leadership in determining how California had gotten itself into this crisis, how the state could get out of the mess, and informing Californians what the damage would be to their pocketbooks, not to mention the impact on their daily lives. Davis's failure to communicate with the public left the people frustrated and wondering whether the crisis was real.

On February 12, a Los Angeles judge denied SCE's request to quickly raise rates by about 9 percent. The judge said SCE's request for a preliminary injunction was totally inappropriate, as it was asking the court to essentially divest the State of substantially all of its regulatory power over rates. The judge did pave the way for SCE to go to trial quickly, accompanied by PG&E, to seek increased rates to cover the more than $11 billion that the utilities claimed they owed generators for power purchases.

Davis had been setting specific deadlines for himself and the legislature for months. In November 2000 he had promised to present a comprehensive plan by December 1 to stave off a full-blown energy crisis, but he ended up offering only a modest slate of proposals. Davis called February 12 a "drop-dead" date, by which he had hoped to cement a deal between the utilities and lawmakers. The date came and went, with no deal. The governor told reporters that he and the state legislative leaders would reach a consensus agreement on a proposal to take to the utilities by February 23, but it didn't happen.

On February 13, the senate committee commenced deliberation on Burton's bill for the purchase of the utilities' portion of 32,300 miles of California transmission facilities. There were some who believed the acquisition wouldn't help the state's energy crisis and would probably create more political headaches than remedies. The State would have to assume the responsibilities of maintaining the system and pay for necessary improvements, which were estimated to be about $1 billion, while the cost of the acquisition was unknown. Gary Ackerman of the WPTF estimated the State cost might be close to $9 billion. Rich Ferguson, former director of the PX, suggested a cost of $7.5 billion. PG&E opposed the proposal outright. SDG&E withheld comment. There was also the potential problem of other buyers, including the wholesalers, wanting to buy the transmission lines. A group in Washington DC expressed an interest in acquiring the grid. Acquisition by the wholesalers would have given them enormous control over the energy market.

About $300 million of a $440 million account financed by the general fund for the DWR to purchase power had been spent by February 15. State finance officials notified lawmakers that they intended to continue spending general fund money—one $500 million increment before February 23 and another increment thereafter. The State planned to start marketing at least a portion of the $10 billion revenue bonds in May and planned to repay the general fund in the current fiscal year. The plan, if successful, would have no significant impact on the current year's state budget.

On February 23, Davis announced he had an "agreement in principle" to buy SCE's transmission facilities for $2.76 billion, coupled with a $420 million contribution from Edison International. Also, SCE would agree to sign ten-year contracts to supply power from some of its power plants. A federal regulator took a dim view of the proposal. A prominent consumer advocate called the plan a giveaway, observing Davis's proposal was a payment 2.3 times the value of the system as carried on SCE's books. Davis contended the plan was "fair and balanced" and would not require rate increases, though he conceded the plan would not be completed until the State was able to acquire both SCE's and PG&E's segments to make a viable package.

However, on the same day he conceded he still had to reach agreements with PG&E and SDG&E for the purchase of their portion of the transmission facilities. The Republican legislature had been opposed to the purchase of the transmission facilities, and some Democrats had also started to question the wisdom of the move. The objective of purchasing the transmission systems was to put cash into the hands of the utilities so they could relieve themselves of the enormous debt they had accumulated.

The plan to purchase the facilities contained a number of pitfalls: (1) the State would need to obtain federal approval, as those facilities were a part of the interstate transmission system; (2) the approval process could take too long to accomplish; (3) the State would not be able to sell revenue bonds to finance the deal until the State had received federal approval; (4) the cost was high—the administration had offered as much as $7 billion for the three utilities' share of the system; and (5) some legislators were of the opinion that the utilities could not be saved by State intervention. If the transmission deal collapsed, Davis reported he had a Plan B that involved buying the utilities' hydroelectric plants. Such an action would provide the utilities with less cash, but would become a source of revenue for the State.

The utilities and state officials in late February conducted a frenzied search for electrical energy and natural gas. With the passage of AB 1x, most wholesalers were willing to do business with the DWR. However, four wholesalers did balk at selling to the ISO. The ISO, which had made only partial payments to suppliers, went to the U.S. district court, forcing the four to keep supplying the ISO. Frustrated with not receiving payment, several small out-of-state suppliers considered refusing further sales. The non-payment burned bridges between California and out-of-state suppliers.

PG&E was encouraged after finding five suppliers—BP Energy Company, Texaco Natural Gas, Dynergy Marketing and Trade, Texaco Canada Petroleum, and Canada Marketing and Trade—that were willing to sell natural gas to them. However, the five could not supply all of the natural gas PG&E required, and PG&E needed to line up additional sources. Two suppliers, J. Aron & Company and Western Gas Resources, planned to cut supplies to PG&E, and those two accounted for 10 percent of PG&E's requirements. PG&E had been current on payment of its invoices, but some suppliers started demanding payment up front, a demand PG&E couldn't meet, which caused PG&E to seek a federal court order for continued supply. PG&E obtained permission from the PUC to effectively guarantee that its gas invoices would be paid with customer payments.

The situation set up a spate of potential litigation between the ISO and four major wholesalers—Reliant Energy, Dynergy, AES Pacific, and Williams—when there were rumblings that some power plants might shut down. However, when faced with a court order, three, excluding Reliant Energy, told the ISO they would continue their supply. The ISO obtained a restraining order requiring Reliant Energy to continue to supply electricity upon its demand. Appearing in court, the four said they were willing to do business with the DWR but objected to being forced to do business with the ISO. The supply was not the question—the question was who would pay.

When California's energy prices were soaring, experts had been asking why the State was paying two to three times the national average for natural gas purchased at the last minute. A Texas company that owned the largest pipeline into California was blamed—it was accused of hoarding supplies that drove prices upward. Other officials cited less sinister causes, stating that the real problem was California's intrastate pipelines and storage facilities. As more data became available, analysts observed that the natural gas fueled power plants producing electrical power had been overlooked. Bill Wood, a natural gas analyst with the CEC, observed that demand helped send prices upward, thus creating a vicious cycle. The cost of one commodity was driving up prices of other commodities. The cycle was triggered by scarcity of hydroelectric power and the financial distress of SCE and PG&E. There was very little help from the neighboring states, causing California's power plants to work overtime and use enormous amounts of natural gas. In November and December 2000, California consumed 2.4 billion cubic feet of natural gas daily; in January the daily consumption was as much as 3 billion cubic feet.

At this time, California was producing only about 15 percent of the gas it required. Southwestern, Rocky Mountain, and Canadian suppliers provided nearly all of the state's natural gas. The greatest supply came into California at Topock, Arizona. In 1992, the federal government deregulated the national distribution of natural gas. Before 1992, there were limited companies that could enjoy the monopoly of buying gas from producers and selling that gas to their customers. Deregulation forced pipeline companies to transport natural gas for anyone able to purchase space in their pipelines, including competing suppliers. The deregulation had helped encourage competition that in turn reduced the price of gas. However, this reduction in prices suppressed any natural gas exploration in the late 1990s, and demand continued to soar while the nation's economy boomed.

Between January and September 2000, spot prices for natural gas doubled to approximately $5 per million BTU at the Louisiana hub, which was the benchmark for the nation. In December, cold weather drove average prices to $9 per million BTU. One week later, California's average spot price leaped to $25 per million BTU. Households and small businesses were sheltered from the gyrating prices to a limited extent because PG&E and other utilities had secured some long-term contracts. Shasta Paper Company in Redding, faced with a $3 million gas bill, shut down its operations and laid off hundreds of workers. Reliant Energy reported it was paying $5 to $15 million per day for natural gas.

On top of it all, the increased requirement for natural gas was exacerbated by the retirement of older, less efficient plants that didn't use natural gas. Over the ensuing three years, various companies were anticipated to build more than twenty sizable power plants in California. Altogether the plants would provide the state with an additional 13,500 Mw of increased natural gas demand.

Natural gas became exceedingly crucial to California's power generation, as new power plants built in the preceding ten years were specifically designed to only burn natural gas. Previously, power plants had been designed to burn either fuel oil or natural gas. Emission standards and the extensive availability of natural gas led the operators to avoid fuel oil.

Prior to deregulation, investor-owned utilities started to remove their dual fuel capability, as fuel oil had become environmentally unsatisfactory and because of the necessity to be in compliance with ARB requirements. At that time, natural gas was plentiful, cheap, and environmentally satisfactory. With the passage of AB 1890 deregulation, utilities were required to divest themselves of their fossil fuel fired power plants. The private generators that purchased the plants asked for and in most cases received permission from the regulators to abandon backup with oil fuel, which was substantially cheaper than natural gas at that time.

California's private utilities had their own large natural gas storage sites. In 1999, Wild Goose Storage was the first private company in the state to offer the storage of natural gas. In the spring of 2000, some electricity generators who depended on natural gas questioned the wisdom of natural gas storage. With rising prices, storage of natural gas risked locking generators into prices that appeared unusually high.

Before the 1990s, utilities used reserve storage for their industrial customers, passing on the cost set by the PUC. In 1992, the state legislature instructed the PUC to create a new system. Large customers like power generators were either to pay the stored prices of natural gas or to rely on the market interstate pipeline price. The purpose of the new system was to open the market to storage companies that were interested in building natural gas storage. Then there were the groups that were dubious. TURN cited a passage written by Niccolo Machiavelli in 1513, "It must be remembered that there is nothing more difficult to plan, more doubtful of success, nor more dangerous to manage than the creation of a new system."

With the power crunch, utilities opted to start tapping their natural gas reserves that were normally reserved for the winter cold weather. Generators could pass their natural gas costs on to the consumer, leading to little incentive to store gas.

As supplies became tight, Californians all needed someone to blame, and the popular target was El Paso Natural Gas Company, which owned the largest pipeline serving California. By law, El Paso was required to sell space in its pipelines to utilities, gas traders, and others. In the spring of 2000 El Paso sold about 40 percent of its capacity to one of its subsidiaries, El Paso Merchant Energy, to the irritation of its competitors. Eventually the price dropped to $13.33 per million BTU, but that price was still twice the national average.

A 1999 study reported that the state needed new power plants that could burn fuel oil in order to guarantee power availability when natural gas became scarce. Thus, the overall implications of the widespread loss of the state's dual-fuel plants became clouded. Elsewhere in the nation, in 1993, the high prices of natural gas caused many duel-fuel plants to switch to burning oil. According to the EIA, as recently as 1993, 43 percent of the state's generating capability burned either natural gas or fuel oil; by 2001, it was less than 5 percent. Some were power plant contracts with the ISO that were essential to maintain California's base load. In 1999, the ISO, pressured by power generators, introduced a study to examine which power plants, if any, should be required to maintain dual-fuel capability. The study identified natural gas bottlenecks, but it did not assess the need for backup fuels to assure reliability of statewide power.

Now that the State was purchasing power on the behalf of the besieged utilities on a long-term basis, the State's problem focused on the staggering debts the utilities had accrued. Governor Davis and the legislature were working on a complicated financing plan in which SCE and PG&E shareholders could repay the billions in debt and avoid additional rate increases. State officials observed that the plan left open the possibility of more rate increases and that it was far from being resolved. Utility executives were adamantly opposed to portions of it, particularly Davis's insistence that the utilities were to surrender "financial instruments" or other assets for the restoration to financial health. The utilities' argument was that they should not be required to give up anything for the use of money used to purchase power for their consumers. What they needed were rate hikes to cover their cost of doing business. The previous month, a federal judge ruled that SCE had the right to pass on reasonable costs to the ratepayers. Davis continued to oppose rate hikes and expected the State to acquire utility assets. Experts opined that bankruptcy could lead to a serious deterioration of the utilities' ability to deliver electricity. Involved in the plan was a concept known as "securitization," where the utilities or the State would issue billions in bonds, and the proceeds would be used to pay the wholesale suppliers' invoices. The bond debt would be repaid over as many as ten years. The plan was designed to disallow rate increases and to dedicate a portion of the revenue to the bondholders. Yet the plan left open the eventuality of rate increases. If the wholesale prices of energy went up extraordinarily high, the utilities would still be obliged to make the bond payments first and then find a way to pay for electrical energy, either by cutting costs or seeking rate hikes. Consumer advocates and many state officials argued that more than \$3 billion, included in the \$12 billion the utilities owed themselves, were profits that they made generating power themselves and argued that the \$3 billion should be deducted from the debt repayment plan.

There was renewed fear that the demand for power could exceed California's energy capability and that the State's nest egg of \$10 billion for long-term power purchases would only be a fraction of the State's potential energy crisis's costs. Accompanying the concerns were (1) the purchase price to acquire the utilities' transmission facilities, which could be as much as \$13 billion; (2) the cost of the below-market energy contracts seized from the PX, which had already been invoiced for \$1 billion; and (3) the cost of improving conservation. Davis had already set aside \$1 billion for energy-related improvements. Consumer advocates observed that either the ratepayers would be hit with rate increases or the power debt would be in the power suppliers' court.

Governmental bodies, including the DOE and some municipal utilities, owned about a quarter of the transmission grid. Whoever owns the transmission grid has the responsibility of maintaining and providing upgrades. The cost of any major expansions, which could run into millions or billions of dollars, would be passed on the consumer. Transmission rates are relatively small and the cost wouldn't have a large impact on electrical rates. However, with the acquisition of the facilities, the major portion of transmission expenses would be paying down the purchase cost.

The plan envisioned leasing the grid back to the utilities to operate and maintain—another cost.

Early on February 27, economists and energy experts, led by David Teece, director of UC Berkeley's Institute of Management Innovation and Organization, issued a manifesto on the electricity crisis, begging the politicians to avoid a political solution. It read, in part:

> Reduce demand; ideally through real-time pricing that reflects the varying cost of power a customer is using at different hours of the day . . . Don't socialize California's electricity system. New government ownership of generation and distribution facilities won't solve the crisis or deliver below-market power prices. The State must pay full market value for any plants it acquires; record electricity prices also mean record purchase prices for energy facilities. For consumers to obtain bargain electricity rates would require permanent taxpayer subsidies, saddling the State budget in perpetuity . . . Don't sign any long-term contracts. Now is precisely the wrong time for the State to commit to long-term contracts for a large portion of California's electricity needs, since below-market prices now can only come at the expense of above-market prices in years to come, which would create new problems. Emergency State contracts should last no more than two years, and should be kept small.

The manifesto was signed by some of the most respected people in the field; among them was Paul Joskow, director of the Massachusetts Institute of Technology's Center for Energy and Environmental Policy Research.

In the past, many advocated the building of the Auburn Dam. In fact, Assemblyman Tim Leslie (R-Tahoe City) and Senator John Doolittle (R-Roseville) had both submitted bills for the dam construction. Republican assemblymen B. T. Collins and David Knowles had both tried. Thomas "Rico" Oller, a Republican representing California's 1st Senate District, remarked, "If the Auburn Dam had been built, we would be in better shape during this energy crisis. We need to build the Auburn Dam now!" A bill was introduced that would have allowed the state to vote on authorizing bonds to fund the construction of Auburn Dam, which would have provided flood control, developed recreational facilities, and provided desperately needed hydroelectric generation. Assemblyman Anthony Pesetti (R-Rancho Cordova), chairman of the Auburn Dam Committee, supported the bill. At the time, the current cost was estimated to be between $500 million and $1.9 billion. Completing the dam would generate between 300 to 800 Mw, though not everybody supported its construction—many environmentalists and rafting groups vehemently opposed the dam, proclaiming it would destroy the river system and would not provide much flood control.

March

Davis's office had announced on March 5 that of the twenty suppliers, Avista Energy was the first to indicate it hadn't reached agreement and was still in negotiations. Reliant Energy also said its negotiations had not gone beyond an agreement-in-concept on its long-term contract. San Diego's Sempra Energy Resources said it had just reached compromises to move forward.

The DWR was the negotiator in the agreements. The long-term contracts that had been announced resulted in an average price of $69 per mwh, $79 the first five years and $61 for the following five years. Davis was confident that was adequate to provide enough cheap power to maintain PG&E's and SCE's existing rate structure. The existing structure included a 20 percent rate increase; the 10 percent emergency rate hike previously approved was expected to become permanent.

State Treasurer Angelides remarked that fixing California's energy crisis hinged on decisions yet to be made. For example, how big of a share of the electricity revenues would be dedicated to the State? Additionally, should the State make sure it would get enough money to repay a $10 billion revenue bond issue covering power costs, with assurances that it didn't get more than its fair share? Angelides claimed the State would need about $1.3 billion a year to repay the $10 billion in bonds.

Under the bill that permitted the State to enter into the power-buying business, a portion of the revenue was supposed to go to the DWR for power purchases; the State would get what was left over after the three IOUs covered its own power generation costs and its QF contract costs. The uncertainty of the formula for allocating money to the State still had to be resolved by the PUC.

Mike Florio, an attorney for TURN, observed that the electricity rates might be higher than expected and would cut into what was available for the State. In his opinion, "The utilities in their initial filings had puffed up their share [of electrical rate structure] beyond reason . . . By PG&E's numbers, there would be nothing left for DWR, which is clearly unreasonable and not what the legislature intended."

Steve Lindsey, a supervisor in the ORA, said, "The biggest uncertainty is how the utilities will be compensated for having to continue to pay for electricity bought on the spot market." The DWR had resisted buying all the high-priced power needed to keep the grid going and depended on the ISO to make up the difference. At the time, the ISO passed those costs on to the utilities, but how the costs would be ultimately resolved still hadn't been determined.

PG&E submitted a letter to the PUC arguing that the continued spot market costs would not only wipe out the State's share, but would contribute even more to the utilities' enormous debt.

Eventually the FERC started to come to the party, ruling that California had been overcharged by $69 million in January. The suppliers were directed to either give back the money or provide detail on how they arrived at the prices they set. The ISO had asked the FERC to order refunds of approximately $315 million for January sales, but the FERC concluded that about 75 percent of that was not justified, as the ISO lacked jurisdiction over the sellers. The $69 million was a big disappointment to many. However, the ISO was encouraged, asserting that generators would think twice before trying to sell at any price. The power generators were obviously disappointed with the ruling and felt confident they could justify their prices and avoid any refunds. The FERC cited thirteen power suppliers that may have overcharged the ISO and the PX. The FERC proposed ending the $150 soft cap on April 30 and reinstating the hard cap, which could be as much as $1,000.

One of California's suppliers, Calpine Corporation, signed several contracts for long-term deliveries totaling $12.9 billion. The contract exceeded the $10 billion legislative package for revenue bonds approved in January. The total power agreements collectively included 8,886 Mw, 7,000 Mw that was to be available in the summer. This represented about 70 percent of the utilities' energy shortfall the ISO had been expecting for the summer. A CEC analyst projected with the additional generation coming on line and with the energy conservation measures, the potential deficit of 5,000 Mw would be offset. While the assembly lawmakers had misgivings about the CEC's projections, Davis remained confident the balance of California's summer energy needs would be covered.

On March 8, the DWR had paid PG&E's trading units $250 per mwh, while at the same time it paid Pinnacle West Capital Corporation $180 per mwh.

On March 14, FERC threatened to order a $10.8 million refund from Williams and AES Corporation, charging that they deliberately jacked up prices by withholding power. Williams argued that the pricing problem had a lot to do with supply and demand issues as well as the cost of natural gas.

On March 14, at 9:09 a.m., the DWR paid Mirant $400 per mwh for power to be delivered the following day, and by 10:00 a.m., the State was paying Mieco $475. BC Hydro had charged up to $1,000, City of Glendale had charged $375, and the City of Eugene, Oregon had charged an average of $450 for its surplus power.

On March 16, FERC ordered six companies to justify $55 million in their February sales to California at potentially inflated prices. The FERC had determined that everything sold at prices over $430 per mwh was excessive and that the ISO was entitled to a $55 million reimbursement. FERC had already on March 9 ordered thirteen suppliers to justify nearly $69 million in potentially overpriced power sales.

The California Legislative Analyst's Office released a study indicating that the state might avoid blackouts during the summer's peak demand. The study anticipated that the hottest days would come late in the summer, that the planned conservation and energy-efficient programs were in place, and that it anticipated new power plants to come on line. In turn, the CEC expected 4,978 Mw from new plants and/or from increased output from existing facilities. The analysis expected a tentative 2,465 Mw by July 1 and an additional 2,513 Mw by the end of the summer. It also anticipated 1,000 Mw of power from peaking plants, although as it developed 250 Mw became a more realistic number.

The ISO estimated the June daily demand shortfall could be as much as 3,030 Mw in May and 6,815 Mw in June. The July shortfall was estimated to drop to 4,685 Mw, and in August 5,297 Mw. These estimates were based on the expectation of a normal California summer. If conservation could accomplish a 10 percent reduction in demand, 5,000 Mw could be saved, leaving a shortfall of 1,800 Mw. Davis projected a combination of new power plants and energy conservation would help the state survive the summer energy needs.

At a news conference, Davis proclaimed new power plants coming on line would produce 5,000 Mw by July and added, "These are megawatts that will stay in California." The CEC issued a memo that chided those pronouncements, maintaining that it would not come close to adding 5,000 Mw of generation by July. It claimed little more than half of that goal was to be on line in July, saying new generation "will come from new base load power plants previously approved by the Energy Commission." The CEC declared that no permits for peaking plants had been issued in the state. Five peaking plants with a capacity 363 Mw were in review with the expectation of licensing in April. Twenty-three other plants were anticipated, but had not been publicly announced. This included thirteen plants with a total capacity of 960 Mw, for which licensing could be expected by September, and another ten with a capacity of 851 Mw to be licensed by the following March. The ISO expected only 500 to 800 Mw to be available by September. The ISO was advised that new power plant building schedules had progressed so little that they were considered canceled. The previous year, the CEC approved construction of 1,300 Mw, while building requests for 3,000 Mw of new capacity had been requested. The reality was that companies interested in building power plants couldn't expeditiously proceed because of all of the federal and state regulations.

FERC said the AES plants operated by Williams had "must-run" contracts with the ISO and required AES to sell power to the ISO at predetermined prices. The contract applied whenever the ISO needed electrical energy. In an ISO order for power from the two AES plants in Orange County, the ISO was told those plants were not available. FERC said the Huntington Beach plant was apparently off line due to improper maintenance. The other plant at Alamitos was said to be unavailable because "Williams took action to extend the outage."

In the summer of 2000, California had been able to import 3,000 Mw daily from the Pacific Northwest. However, the current summer was another story; the Northwest experienced a drought of major proportions. It was questionable if the Pacific Northwest would have sufficient power to meet its own requirements, much less those of California. The Bonneville Power Association markets power from twenty-nine federal dams, mainly in Washington and Oregon, and supplies 40 percent of the region's power. As the drought deepened, the BPA and the region's utilities were without adequate power to meet their own requirements and had to purchase power on the spot market. The snow pack was 53 percent of normal. Alarmingly, if the Grand Coulee Dam water level fell another 15 feet, enough water to generate 6,000 Mw, it would not be able to generate any power. Environmentalists accused California of killing salmon by taking water from Washington and Oregon to generate electricity for export to California. Eight times the BPA invoked a provision of the Endangered Species Act, which allowed it to generate electricity instead of storing water for salmon runs. In normal years the BPA sent power to California during the day and at night imported power from California, reducing the production of electricity and thereby saving water for the salmon runs. The BPA had been making millions of dollars selling power to California's utilities and other power purchasers. With the shortage of water, the BPA ran up a $500 billion bill from purchasing power on the spot market.

On March 20, blackouts struck about a million Californians, and DWR paid $565 per mwh for energy.

On March 23, PG&E threatened a lawsuit against the State if the state legislature passed several measures that the legislature believed were necessary for controlling California's electrical energy efforts. In a statement given to reporters, PG&E said, "The proposed actions will destabilize [PG&E's] current situation with creditors and put the entire possibility of reaching a balanced solution to California at risk." At the same time, assembly members were divided on how to keep hundreds of alternative energy providers alive and assure the repayment to the State of billions dollars.

Then PG&E pulled a faux pas—PG&E's parent company disclosed a $1 billion refinancing loan from Lehman Bros. and GE Capital Structure. The loan was to pay $501 million to bond holders, $434 million to miscellaneous lenders, and $116 million to pay the delinquent year 2000 fourth-quarter dividend to the shareholders. The terms of the loan required PG&E to pay the fourth-quarter dividend because it was considered a debt in default, but it forbad the payment of future dividends while the loan was in effect, which might last two to three years. A PG&E bondholder suggested that with the loan, the corporation might strengthen its bargaining power with state officials by erasing the threat of a parent-company bankruptcy. Assemblyman Dave Cox (R-Fair Oaks) wondered if the deal would impair the State's ability to purchase the transmission lines, as the company would now clearly be creditworthy. Davis faulted PG&E for cutting a complicated restructuring deal to isolate PG&E's National Energy Group from the utility problems. Others argued the deal "provided more fuel" for those who argued that the parent company should put more money into the beleaguered utility. Wall Street Fitch Investors Service admonished the PG&E parent company for not having taken similar moves to refinance the utilities debt, "Failure to promptly resolve the utility subsidiary could make bankruptcy a near possibility." Reliant was annoyed with PG&E and SCE for lack of payment on a $300 million debt for power already delivered. Reliant vice president Jack Strout wondered, "If the procurement of their transmission system pays them $2 or $3 billion, what are they going to use that for?"

Governor Davis remained adamant that he could hold the line on the existing rate structure and seemed to ignore what many advisors had been telling him. The *Sacramento Bee* on March 25, 2001, had this to say:

> John Burton, Senate President pro Tem (D-San Francisco), said, "I think the facts finally came out in terms of the magnitude of the problem . . . You have to get money from somewhere—unless you steal it or win a lottery ticket . . . I don't know [how] to solve this problem without [a rate increase] and I never saw how you could solve it without one." Key lawmakers were told by Davis's staffers that the State may need to borrow as much as $23 billion, and raise rates by as much as 60 percent to 80 percent above the initial 19 percent to support energy requirements over the next eighteen to twenty-four months . . . "It's insane and disgusting at this point," declared Doug Heller, FTCR. "Governor Davis is unfortunately tending to his campaign coffers more effectively than tending to the state's energy needs." . . . After his briefing with the administration, Assembly Speaker Robert Hertzberg, (D-Sherman Oaks) said, ". . . it seems more inevitable that the Public Utilities Commission will have to raise rates." Angelides said that the numbers suggest the State can't survive financially under the current rate structure. "There's a limit to how much you want to sell bonds to keep delaying reality," he said. "At the end of the day the PUC is going to have to set rates that allow the State to buy power and the State to recover their costs . . . How many billions are we willing to borrow to delay the tough reality of where we are?"

On March 26, a proposal was revealed that would have raised electrical rates approximately 30 percent, the biggest rate increase in the state's history. Minimum users would be let off sparingly, while the biggest users could experience rate increases up to 90 percent. Governor Davis distanced himself from the proposal. The PUC released a package of proposals for possible vote action the same day, including suggestions that legislators (1) make permanent the temporary 9 percent, 1¢ per kwh rate increase that was imposed last January; (2) start immediately designing rates that would hit the heaviest users the hardest; (3) revise the formula for payment to alternative power generators in a way that would solve the smoldering payment dispute; (4) create a formula for the utilities to pay the State for electrical energy purchases and clear the way for the State to issue power purchase revenue bonds; and (5) reclassify past debts and expenses to sharply reduce the utilities' estimates of their losses.

While investors embraced the proposals, consumer groups belittled them, and the utilities were cautious, but economists and generators labeled it as a good start. Lynch said it was "time to pay the power bills for California."

The proposal would have allowed PG&E to collect an extra $2.3 billion annually. PG&E's average rate would increase from 1.04¢ to 1.34¢ per kwh, bringing the total rate increases for the year to more than 40 percent. Individual consumers, especially low-income and residential consumers who were below a state-mandated poverty threshold, could conceivably avoid any rate increases. Companies big enough to operate their own substations could face an 87 percent increase, agriculture consumers 10 percent to 24 percent, and other businesses 41 to 58 percent. Lynch said, "Electricity hogs will have to pay more." Davis told reporters, "As governor, I have not decided there should be a rate increase, and as governor, I have not decided that tiered pricing makes sense." Michael Shames, head of UCAN, said, "This PUC doesn't do anything of substance without the governor's approval." Lynch believed the new rate increases would be enough to pay the QFs and the State for future power production. However, it did not resolve what was needed for payment of past purchases that had driven the utilities into serious debt. Dean Vansch, president of Delta Power Company, was pessimistic about the proposed plan. Delta had five gas fired cogeneration plants, of which four were shut down for lack of payment. Utility stocks went up nearly 30 percent. Lynch relied on the State bond offering to finance most of the DWR's purchases of electricity.

On March 27, the PUC approved the proposed rate hike amounting to $4.8 billion a year but wouldn't guarantee the new rates would solve California's energy crisis. Major issues remained unresolved. The increased rates didn't pay off the enormous utility debts. The utilities were still subject to possible bankruptcy fillings by power suppliers, bondholders, and other creditors. The increases might promote conservation but might not be enough to prevent rolling blackouts. The rate hikes might not pacify alternative energy producers, which provided 20 percent of California's energy requirements. Hundreds of QF generating units had been shut down for lack of payment. The increased revenue would primarily be allocated to pay for energy the DWR's purchased. Governor Davis steadfastly remained in opposition to any rates hikes.

The PUC's rate hike made utility bankruptcy more tolerable to state officials, as it insured payment for ongoing power supplies. The two major utilities criticized the PUC for peremptorily lowering the estimates of utility debts. The PUC estimated the increased rates would generate additional yearly revenue for the two utilities—$2.28 billion for PG&E and $2.51 billion for SCE. Gary Ackerman of WPTF observed, "You still have $14 billion, maybe more, of costs that haven't been paid for, and that's still a problem. I don't believe creditors are going to be indefinitely patient."

The PUC's rate hike represented an exceedingly hard hit for large businesses. The highest rates would at occur at peak times when businesses were their busiest; their rates would see an overwhelming increase of 87 percent. Small commercial businesses would see a relatively modest increase of 16 percent, and residential customers would see an increase of 24 percent.

Two assembly committees met March 28 in closed meetings to explore the causes of the previous winter's soaring natural gas prices. Davis met privately with Democratic assembly leaders, reassuring them that his electricity plan was still on track. The energy agenda dominated the committee meetings. The FERC dismissed PUC accusations that El Paso Natural Gas Company rigged the bidding for California pipeline capacity to favor a subsidiary. The FERC did censor the parent company, El Paso Corporation, for conniving to use its power to drive up natural gas prices in California. In a two-hour meeting, Davis reviewed the progress made to enable the State to purchase a considerable portion of California's transmission facilities. State Controller Connell warned that the newly approved rate increase would not cover all of the State's power purchases. She projected that the State budget would be left with more than a $7 billion cash deficit over the next year and half.

State Treasurer Angelides cautioned that any legal action could delay his efforts to sell bonds. Of particular concern was the repayment of the amount in excess of the $4 billion the State had already spent on electricity, which the State had purchased on behalf of the utilities. The legislature had to address paving the way for marketing the $10 billion in bonds to finance the long-term energy contracts. The Democratic legislature proposed inserting language that would make repayment to the State a priority as the utility customers paid their bills. Opponents argued the issue was not quite that simple. In a worst case scenario, the State and the bondholders would get payment first and leave an inadequate amount for the utilities. At that point, hundreds of alternative energy providers might not have received payment. If they were not paid, they threatened to take the utilities into bankruptcy, unless of course, they could sell their power to the State. Davis proposed an agreement that would guarantee payment for any power produced from then on. Collectively, the utilities owed the alternative energy producers $1 billion. Current electrical rates were tied to the price of natural gas. If one-price-fit-all, then alternative energy providers would find themselves being paid far less for their energy than the cost they needed to produce it. To exacerbate the problem, the alternative energy producers feared they would not be treated fairly by the PUC.

California lawmakers debated a piece of legislation designed to reduce the prices that alternative energy suppliers charged. Assembly Speaker Pro Tem Fred Keeley worked with his senate counterparts to hash out differences. He warned lawmakers they might not be able to craft a total package that would keep electric rates below an expected 19 percent rate hike next year. "Events have unfolded during the past month or so which make it very unlikely this overall financial situation can be stabilized within the existing rates." Keeley said he would work with Senator Debra Bowen (D-Marina del Rey) to craft a new bill with the objective to meet a price of 7.8¢ per kwh for energy produced by alternative energy generators. Speaker Hertzberg united with several legislature leaders who believed a rate hike was justified. Utility officials maintained the proposed rates were simply not high enough to cover both the State's requirements and their own costs as well. One leader suggested a bond sale of $23 billion might be necessary. Davis remained unconvinced and argued they should continue to work within the existing rate structure.

A number of experts believed Davis had been hasty in his long-term deals. They observed that while it was prudent to reduce California's reliance on spot market rates, California should have followed neighboring states' examples and signed contracts lasting only three to four years. An opinion was expressed that the contract rates would certainly be lower than the current spot market, but once the crisis was over, with additional generating capacity on line, the State with shorter term contracts would be more flexible in its ability to bring down costs over the long haul.

After the PUC meeting on March 29, PG&E president and CEO Gordon R. Smith weighed in:

> "Today's CPUC actions, while providing a welcome dose of realism in the energy policy debate, unfortunately do not resolve any of the key issues facing California's policy makers during the energy crisis. The actions do not offer a comprehensive solution, fail to resolve the uncertainty of the crisis, and may even create instability. Finally, the commission has included a number of elements in its decision and are, in some cases, illegal attempts to change rules after the fact. For example, the commission has ordered the company to make payments retroactively to the Department of Water Resources' purchases. This order mandates a payment that upsets the company's efforts to treat all creditors in a non-discriminatory manner . . . Most egregious, however, was the attempt by the commission to completely change the rate making rules that are to determine the end of the rate freeze under AB 1890, rather than acknowledge utility arguments that the rate freeze have been met, the commission attempts to order the recovery of stranded costs, going back to 1998, and then to account for those costs, in a manner contrary to law. There is no justification for this action, other than to artificially extend the rate freeze. Pacific Gas and Electric Company will challenge this and other aspects of the decision."

Utility executives said the PUC ordered all the revenue from the 30 percent rate increase to be used to repay the State for the DWR power purchases. The utilities argued that the PUC's order would hinder their ability to pay for energy from the alternative energy suppliers, such as the wind energy that accounted for 20 percent of PG&E's supply. SCE officials questioned that even with the additional money coming in, how could it make the State, the utilities, and the QFs whole? The PUC, in its authorization of the rate hikes, ordered the utilities to begin paying the QFs for their power deliveries starting in April.

PG&E asked FERC to increase the Western Area Power Administration (WAPA) rates for power it was transmitting from its Central Valley Project. WAPA is a federal agency for marketing power throughout the West. The utility said if the rate was approved, the rates it charged SMUD could increase as much as 20 percent. PG&E claimed it had been forced to go to the open markets to firm up WAPA deliveries, which added $383 million to PG&E's costs between April 1, 1998, and January 1, 2001. It added that its costs would increase by another $1.2 billion before December 31, 2004, when the contract expired.

The PUC "passed the buck" to Governor Davis, stating that all the money the DWR was spending on power could be passed on in electric rates in one of two ways, either by wrapping it into existing rates or by increasing rates. The decision essentially told the governor he could direct the PUC to impose whatever rates were necessary to repay the DWR for buying power on the behalf of the cash-strapped utilities.

In an effort to conserve cash, PG&E and SCE petitioned the PUC for a reduction in force with the intention of laying off 3,000 employees. It prompted a union opposing petition. The PUC voted four to one to rescind the layoffs; the utilities were directed to rehire the 725 employees previously laid off and to scrap any plans to lay off an additional 2,225 employees. The commissioners stated they were acting to preserve reliable service for the consumers. PUC president Lynch argued that the layoffs would lead to lengthy delays in new service installations, meter readings, and other elements of basic services. PG&E estimated that laying off 1,000 workers could save $360 million a year. SCE in turn indicated that laying off two thousand workers was part of a $465 million a year cost-reduction program and said that the PUC decision only served to exacerbate its financial crisis. One PUC Commissioner cast a dissenting vote, saying, "The commission decision should not have interfered with the utilities' efforts to economize."

During the month of March, a number of other issues were addressed. The legislature was holding up SB 47x, a bill addressing the cost of power from 688 qualifying facilities, low polluting alternative energy producers that contracted directly with IOUs. These facilities were the outgrowth of previous legislation designed to encourage the development of alternative energy sources such as solar, wind, and cogeneration. The bill was popular in the legislature and the politicians looked to sandwich its provions into future measures such as a bill, yet to be designed, that would give some financial life to debt-ridden utilities. The current price for this power was tied to the skyrocketing price of natural gas. If QF power could be switched to long-term contracts, the price of gas could be halved; however, it required the legislature to rewrite the formula that would lock the power into long-term, five-year contracts.

Senate bill SB 5x was rejected; the bill would have spent $1 billion on energy conservation. Davis was optimistic that the bill was not dead, and he expected it to be passed soon.

With deregulation, the ISO had been put in control of transmitting power over PG&E's transmission lines, and PG&E claimed in its filing that transmission produced $78 million in unforeseen costs that could swell to $191 million before the contract expired.

The City of Roseville and Enron reached a mutual agreement for the construction of a 750 Mw natural gas fired steam power plant to service South Placer County communities. The plant would be built on twenty-two acres next to Roseville's pending wastewater treatment facility. As owner, Enron would build and operate the plant in exchange for use of the land and use of the wastewater from the treatment facility for cooling water. A part of this agreement benefiting Roseville would be the sale of the treated water to Enron. These additional funds would allow South Placer sewer rates to remain low. The communities would also realize monetary benefits—Roseville would receive annual development payments of $25 million for twenty-five years after the plant was operational. Construction was estimated to take four to five years. South Placer communities would also receive additional annual monetary benefits: Roseville, Rocklin, and Lincoln, each $200,000; Placer County, $250,000; and Loomis, $150,000. The project was estimated to produce 300 construction jobs, twenty permanent operational jobs, and $205 million in payroll, sales taxes, and material purchases. Enron anticipated filing an application with the CEC in June.

Governor Davis, using his emergency powers, announced his 20/20 conservation plan, which would offer a 20 percent rebate to customers who conserved 20 percent of their normal energy use during the hot summer months. The benchmark for determining the conservation rebate would be the previous year's use. Davis offered about $90 million in rebates for the reduction of energy use. Interruptible customers were required to shave 20 percent during peak hours in order to be entitled to the 20 percent savings. John Nelson of PG&E calculated that the average customer who used 20 percent less energy over four months would receive a rebate of approximately $42. That rebate was in addition to the savings the customer would realize from using less energy. The rebates would show up as credits on each customer's monthly bill.

April

The threat of blackouts was not limited to California—all the western states participated in a "club membership" they had not opted for. The Pacific Northwest, for example, experienced electricity rates higher than California's. Robert Dintelman of WSCC proclaimed, "I don't recall a time when it was as tight as this." University of California Energy Institute's Severin Borenstein was quoted as saying, "The western grid could be very short and when that happens, the price goes through the roof . . . we get a scramble for power." California's power problem, however, was at the head of the line because it imported 20 percent of its energy requirements. But prices had escalated in Oregon and Washington as a result of a drought. The previous April, wholesale power at peak times sold at $36 per mwh in California, $31 per mwh at a Washington trading hub, and $35 at an Oregon hub. On April 3, 2001, the peak prices in California were $300 per mwh, in Oregon, $345, and $350 in Washington. Duke Energy spokesman Tom Williams warned, "I don't think imports are going to show up in California this summer. And if they do, they're going to be very, very expensive."

The PUC's recent rate increase did not make many happy. Alternative generators (QFs) were planning to do battle with utilities and the PUC. The QFs called the new payment formula inadequate and asked to be released from their contracts with the utilities. Three QFs sued SCE for release. Watson Cogeneration Company, a plaintiff, one of the largest generators serving the BP Amoco oil refinery as well as approximately 400,000 residences in the Los Angeles area, was owed $150 million by SCE. In an earlier court ruling, CalEnergy Company had been released from its contract with SCE. This encouraged other QFs to sue for release from their contracts. Jerry Bloom of the California Cogeneration Council predicted, "You're going to see a slew of lawsuits." He went on to say that companies that couldn't get out of their contracts were going to be faced with a difficult time producing power at the rates set by the PUC. Hundreds had terminated their production of power for lack of payment by PG&E. Those QF shutdowns deprived California of 3,000 Mw of power and contributed to the rolling blackouts in March.

If the QFs became successful in getting free of their utility contracts, they could help supply power to the grid, but at higher prices. If they sold their power within California, they would be able to sell it on the open spot market, making considerably more money than under their utility contracts. With their newfound freedom, they could also sell their power out of state.

The QFs were not the only problem facing policy makers. The PUC pointed out that with its rate change, it didn't direct the utilities to send all of their additional money to the DWR and that they had left enough money to pay the QFs for their power. QFs running on natural gas were especially agitated with the PUC, which had tied their rates to the price of natural gas at the Oregon border, a price the QFs claimed was far below the true market price, particularly in Southern California.

On April 5, Reliant Energy asked the federal appellate court to overturn the lower court ruling that Reliant must sell power to the State in an emergency without assurance of payment. The U.S. Court of Appeals for the Ninth Circuit granted an emergency stay, saying Reliant had shown "a high rate of success on the merits." The ISO said the decision left the State in a more fragile position, as it gave no recourse in the event Reliant shut down any of its plants for maintenance. A lawsuit was brought on by the ISO to force the sale of electricity by Reliant Energy, Dynergy Power Corporation, AES Corporation, and AES's marketer, Williams Energy. The last three defendants had entered in a written agreement to continue to supply emergency power until the FERC decided whether they were required to sell to unworthy companies. In the meantime, a QF supplier won a court order to dodge the two utilities and sell its power on the open market.

SCE and PG&E each took different solutions to resolve their enormous debt. SCE negotiated a contract with Governor Davis, while PG&E elected to pursue its debt resolution through the bankruptcy court. Mitch

Wilk, consultant and former PUC president commented, "We now have two models; it will be interesting to see who recovers the quickest." PG&E was hit the hardest, accumulating a debt of $6.6 billion, while SCE accumulated a debt of $3.5 billion. Davis made an offer to PG&E that would have paid off its debts; it included $3 to $4 billion to purchase PG&E's share of the transmission grid. AB 1890 had called for the cap on electricity retail rates to end in 2002, but Davis wanted to extend the cap for ten years as part of the deal. PG&E wanted an immediate end to PUC regulation and would agree only to a cursory review of rates through "a jury-rigged arbitration," according to Davis. PG&E said it wasn't seeking an end to regulation; what it wanted, which Davis's negotiator Michael Peevey agreed to in principle, was to let the utility ratepayers pay for the full wholesale price of electricity. It was rumored PG&E had a gentlemen's agreement to sell its portion of the transmission grid to the State at an unspecified price, and several weeks later the Davis administration reneged on any agreement. PG&E had become so disgusted with a series of PUC rulings that it claimed the rulings hindered its ability to recover its debts. The issue came to a head on April 5. In a televised address on that date, Davis unhappily came to the rate party and announced a three-tiered rate increase. He said, "Simply stated, we have two problems—supply [is] too low and costs are too high . . . both result from the flawed deregulation scheme created back in 1996." He pointed out that in the twelve years before he took office, not a single major power plant had been built. He also cited the fact that since he become governor, twelve major power plants had been licensed and ten more were in the pipeline. He outlined seven steps he had taken or planned to take in order to "keep power on and the economy strong": (1) he had negotiated long-term contracts at vastly lower prices; (2) he had used his emergency powers to seize control of low-cost contracts that the utilities were about to surrender to generators; (3) he began negotiations to purchase the utilities' transmission systems; (4) he cut red tape and offered cash incentives to speed up the construction of power plants; (5) he was launching an $800 million energy conservation program; (6) he was moving to establish a public utility authority to enable the State to build power plants if the private sector failed to keep ahead of the state's power requirements; and (7) he said the State was using every possible legal remedy to root out and punish illegal conduct.

In the tiered residential rate increase, usage below baseline consumption would see no change in rates; usage 130 percent to 200 percent of baseline would see an 11 percent increase; and usage above 200 percent would see a 37 percent increase. Davis stressed that the more a household used energy, the more it would pay, and the more it conserved, the less it would pay.

Since mid-January, the State had committed $5.7 billion from the general fund to purchase electricity for the beleaguered utilities. The commitment ate into a significant portion of the State's surplus. State Treasurer Angelides and others had estimated the current surplus to be between $6 billion to $8 billion. Angelides was anxious to get the bond bill passed by May 8, as any delay jeopardized his bridge loan of $4.1 billion; lenders were hesitant to make the loan until they were assured the bonds would be sold. The downgrade was estimated to add $50 million to $100 million to the borrowing costs over the life of the revenue bonds.

Angelides had assembled a package in $4.1 billion of loans to pay for energy until the sale of revenue bonds was accomplished. He hoped to increase the package to $5 or $6 billion with bridge loans. The loans due August 29 would be paid off with the bond money; however, he stated that any delay in accomplishing the bridge loans could jeopardize the repayment of the loans. Before the bonds could be sold, he would need more information on the DWR's long-term contracts that Davis had been keeping close to his vest. Tim Gage, Davis's finance director, announced that Angelides's bond underwriter, J. P. Morgan, had suggested more than $10 billion could be needed.

The following morning PG&E filed for bankruptcy, dismissing Davis's speech as "a lot of words." Nancy Day, a consultant for the legislature, believed Davis's PUC members would "punish" PG&E for its behavior, while if SCE filed for bankruptcy, it wouldn't be punished. Plunged into enormous debt and frustrated by ineffectual negotiations with the State, PG&E filed for Chapter 11 bankruptcy on April 6th. The bankruptcy protection proceedings would allow PG&E to continue doing business, but under bankruptcy reorganization rules. Many of its decisions would have to be approved by the presiding U.S. bankruptcy judge, Dennis James Montali. The court case was likely to take years to resolve the issues, and the cost might be $1 billion before PG&E would be independent of bankruptcy. A Chapter 11 filing would allow PG&E to keep creditors at bay and enable the utility to reorganize its finances. PG&E would continue to retain both its utility operations and the management of its business operations. There would be the question of who would ultimately get the utility's assets, PG&E's shareholders or its creditors? The fundamental obligation of the bankruptcy process was the repaying of the utility's creditors, and in that process not everyone would be treated equally. Those creditors holding collateral had to be paid before the payment of lower creditors, vendors, and other lenders not holding collateral.

A big uncertainty in the utility's bankruptcy filing was the issue of rates. For example, did the judge have the authority to increase rates to meet the creditors' demands? Some experts believed a bankruptcy judge does not have that authority. An attorney representing bankruptcy creditor in Tucson, Arizona opined, "There's been a lot [of] academic speculation on both sides, but it's never been settled." Utility bankruptcy is a rarity, but among those that have succeeded were the Public Service Company of New Hampshire, Tucson Electric, and El Paso Electric, but precedents in those cases didn't contribute much to this issue. Another question had to be answered, where does the DWR appear priority-wise on the list of creditors?

There was uncertainty about how alternative generators fit into the picture. Many that remained unpaid had shut down their facilities. PG&E took the stance that it might pay generators only for future delivery and only if it could afford to, saying, "We do not intend to pay out more than we collect in rates." PG&E was amenable to releasing alternative energy suppliers from their contracts, freeing them to sell to anyone willing to purchase their energy.

With the bankruptcy filing, the State's influence over what would happen was now diminished. State Treasurer Angelides said he was comfortable with the filing. He observed that the filing could hinder the big power supplier's ability to get paid. The energy producers were happy, as the filing removed the issue from the politicians' influence.

In the filing, PG&E listed assets of approximately $24 billion and debts of $18 billion. The top unsecured creditors are listed in the table below.

Creditor	**Indebtedness**
Banks, bond mutual funds and others	$4,448,011,000
California PX	$1,996,000,000
California ISO	$1,128,800,000
Power generators:	
Calpine Corporation	$174,663,880
Crocket Cogen	$48,400,372
El Paso Merchant Energy Gas LP	$40,147,245
GWF Power Systems LP	$40,122,073
Geysers Power Company LP	$32,867,878
BP Energy Company	$29,523,530
Enron	$28,210,551
Chevron	$24,716,334
Sempra Energy Trading Corporation	$23,849,455
Wheelabrator	$21,508,087
Sierra Pacific Industries	$19,888,348
Total	**$8,056,708,753**

As PG&E's bankruptcy filing had placed the issue in the hands of the bankruptcy judge, it drew substantial comment from lawmakers and others. Senator Burton said, "Bankruptcy court is the fairest, most stable venue for dealing back debt issue. [Now] there's no need for the State to take steps toward a bailout . . . Let them work it out at their own pace in the venue they chose."

State Treasurer Angelides said the Chapter 11 filing limited the State's options, but in the long run it increased the prospects for a State public power authority. "I've always thought it was odd we should be so anxious to have the ratepayers make sure the utilities got enough money to pay generators 100 percent of their outrageous prices." The filing "does reinforce the notion that you're always better to control your own destiny . . . This whole situation should remind people that we need to review energy as a necessity, not just as a commodity to be traded on some marketer's screen each day. We're assessing, with our attorneys, what impact, if any the filing will have on our ability to get the bridge loan done. Maybe there's been a couple more hurdles on the track, maybe not. Our job is to clear them."

Moody's Investor's Services said the action "increased the risks to the State's otherwise strong fiscal and economic condition."

California Attorney General Bill Lockyer asked the public to assist in his investigation into whether energy generators were guilty of illegally manipulating the gas and electrical markets. "Since billions of dollars may be recovered, the award to an informant could potentially range from $50 million to hundreds of millions of dollars."

Lawyers for Davis and Lockyer met to determine the State's next legal move. Davis's spokesman Steve Maviglio was to seek a court order requiring PG&E to return to the bargaining table. "We always want to work across the table, but we aren't willing to let the utilities to run roughshod over the ratepayers . . . I believe PG&E is petitioning for bankruptcy to force an expedited negotiation that will come out in their favor. They are not satisfied with the deals that have been offered so far, so they're trying to ratchet up the consequences." Maviglio conceded that negotiations between the State and PG&E were pretty much at an end.

Harvey Rosenfield of FTCR said the State's work to date on the transmission line deal and proposed rate changes were virtually wiped out, and it would have to begin again in the legal arena. He acknowledged that in the long run, bankruptcy might be better for both the utilities and the consumer. "It's not going to be subject to political shenanigans. It's not something that's going to be kicked around in the legislature behind closed doors with all sorts of numbers that don't add up. It's going to be done in a courtroom with representatives of all interested parties. We're going to see the numbers, and we're all going to know what is going on. Bankruptcy is a process that is designed to bring order and protection to companies that run into trouble. There is bankruptcy so there is not chaos—so there is not what we've had in the last four weeks."

PG&E was in bankruptcy court on April 9th requesting more time to prepare revised accounting documents for the PUC. PUC's lawyer Gary Cohen argued the bankruptcy judge did not have jurisdiction over the request, saying, "This is the first volley in PG&E's legal strategy, which is to try to get the bankruptcy judge to subvert the PUC's regulatory authority." A PG&E spokesman countered, "It isn't true; it isn't what happened today." The energy suppliers had hoped the judge would ultimately order or pressure the PUC to raise electrical rates.

Robert D. Glynn, Jr., chairman of PG&E, said:

"We chose to file for Chapter 11 reorganization affirmatively because we expect the court will provide a venue needed to reach a solution, which thus far the State and the State's regulators have been unable to achieve . . . PG&E decided to file for protection under Chapter 11, April 6, primarily due to (1) Failure by the State to assume full procurement responsibility for PG&E's 'net open position,' as was provided under AB 1x. This was the result of increasing financial exposure to un-reimbursed wholesale energy procurement costs, which the utility estimated to be $300 million or more per month; (2) The impact of the inaction by the PUC on March 27 and April 3 created new payment obligations for the company and had undermined its ability to return to financial viability; and (3) Lack of progress in negotiations with the State to provide recovery of $9 billion in wholesale power purchases made by the utility since June 2000, which have not been recoverable in frozen rates. The adoption by the PUC of an illegal and retroactive accounting change that would appear to eliminate our true uncollectible wholesale costs . . . Despite Pacific Gas and Electric's best efforts to work with the State of California to reach a consensual, responsible, fair, and comprehensive solution to California's energy crisis, no agreement has been reached with the governor and the governor's representatives have dramatically slowed the pace and the progress of discussions over the past month. Furthermore, since last fall, we have filed comprehensive plans for resolving this matter with the CPUC, but they have not acted affirmatively on them. In November 2000, we filed our rate stabilization plan, which, if adopted, would have increased rates by an initial 25 percent, compared with the 46 percent recently adopted by the CPUC. Neither request was acted upon . . . Had the State acted at the time, Pacific Gas and Electric would have been kept creditworthy; Pacific Gas and Electric would have been able to enter into long-term power purchase contracts at prices lower than those announced by the State; the State would not have had to almost exhaust the State's budget surplus by spending billions of dollars to purchase power for the utility's customers; the State would not need to issue billions of dollars to cover these power purchases; and the State would not now be advancing a proposal to spend billions of dollars to purchase the state's three investor owned utility's electric transmission systems . . . This year, the State has spent more than $3 billion on power purchases, and with the CPUC, has arranged to be reimbursed for these expenses. In contrast, since June, Pacific Gas and Electric Company has spent $9 billion in excess of revenues to pay for power for its customers and exhausted its ability to continue borrowing, but there as been no progress on a plan to reimburse it for those expenditures as provided by law . . . Statements by the governor and other public officials since last September gave us reason to believe a solution could be reached outside the context of Chapter 11 that would restore the utility's financial viability and enable it to meet its financial obligations equitably. However, these statements have not been followed up by constructive actions, and reorganization in Chapter 11 is now the most feasible means of resolution."

During the Chapter 11 process, Glynn assured the public that (1) the utility would utilize its existing resources to continue operating its business during bankruptcy, including paying vendors and suppliers in full for goods and services received after the filing; (2) the utility would pay electric commodity suppliers as provided by law; (3) the utility intended to continue normal electric and gas transmission and distribution functions; (4) employees would continue to be paid, and health care plans and other benefits for employees and most retirees would continue; and (5) the utility's qualified retirement plans for retirees and vested employees were fully funded and protected by federal law.

On April 9, three days after PG&E filed for bankruptcy, Davis announced an agreement with SCE to buy SCE's portion of the transmission system. Under this agreement, the SCE was obliged to (1) pay off $1.3 billion of its debt from the sale, with the balance paid with corporate bonds to be repaid through a portion of its electrical rates; (2) sell its portion of the transmission system for $2.76 billion; (3) use its generation assets to provide low-cost regulated power to the state for ten years; (4) dismiss lawsuits for higher electrical rates; (5) commit the entire output of generating facilities to providing energy to California at a fixed price for ten years; (6) grant perpetual conservation assessments—20,600 acres of the lands related to the Big Creek hydroelectric plant and 825 acres of the eastern Sierra; (7) grant perpetual conservation easements on more than 21,000 acres of forestland; and (8) invest $3 billion in capital improvements in its distribution system, to be covered by a rate increase yet to be determined. In addition, Edison's parent company was to refund $400 million to SCE.

The State of California, for its part, was obligated to (1) continue buying power through December 31, 2002, (2) allow SCE to issue $3.5 billion in bonds for debt not covered by the $1.5 billion SCE would realize from the sale of the transmission system, and (3) allow SCE to continue receiving an 11.6 percent rate of return for ten years.

Governor Davis supported legislation SB 6x that would create a public power authority to construct power plants and help finance conservation and renewable-energy projects. Republican legislators opposed Davis's plan to purchase the transmission systems and believed the State would be making a big mistake by seizing private plants and/or having the State build power plants. Assembly leader Dave Cox said, "There's no reason to believe the State of California can run the power business or transmission line as good as, or better than, the private sector. How far do we take this? If we have a natural gas shortage, do we then seize those pipelines?"

Jan Smutney-Jones, executive director of Independent Energy Producers Association, said the private companies were investing billions to generate power and were running their existing plants as hard as possible to meet the energy crisis. Duke spokesman Tom Williams said Duke was spending $1.6 billion to upgrade and expand its facilities in an effort to ease the energy crunch and State seizures would only discourage and hurt the industry: "The governor has said two things need to happen—the state needs more supply and lower prices. Seizing power plants would do nothing to help those two objectives. We would fight it vigorously, to ensure we got fair market value. We support any effort, public or private, to add generation to California. The state needs it—and needs it fast."

With a bill moving through the legislature to create the State power authority, some municipalities expressed renewed interest in forming their own power authority. They were watching what would happen in the cities of San Francisco, Corona, San Diego, and others in that event. The cities of Davis, Lincoln, and Culver City had already looked at the possibility of forming their own power authorities and had decided against it. In November, the city of San Francisco was scheduled to vote on taking over the city's electrical system. PG&E opposed the referendum—PG&E spokesman John Nelson said, "We will take it very seriously." He went on, saying, "There's been much talk about it, but as soon as most folks sit down and take a look at the realities . . . They realize they have to buy [energy] on the same market, at the same outrageous prices that we and the State have been paying. The talk quickly evaporates." PG&E argued that the factors that once made public power attractive, such as access to low-cost electricity from federal dams, had disappeared. Additionally, if the cities were successful in withdrawing from PG&E and SCE, they would have to allocate money for years to come to pay off the still unsold bonds they would have to assume, thereby adding to power costs for up to fifteen years. The cities argued they could charge less since they didn't have to pay dividends to investors or pay taxes on their sales. What they didn't seem to recognize was that they would be losing revenue from property taxes.

With the high potential for frequent rolling blackouts, many leaders were getting concerned about industries using emergency diesel engine electrical generating units. This was of particular concern during the summer months when there was a wide increase in the use of air conditioning units. Air districts reported there were 641 diesel units in the Sacramento area, 5,000 units in the Bay Area, 1,200 in the San Joaquin Valley, and thousands more in Southern California, resulting in smog plaguing the valleys. Most of the units were scattered throughout the state at hospitals, universities, factories, and businesses. Most of these generating units were grandfathered in the current pollution regulations and were anticipated to be used only during emergencies such as earthquakes and severe storms. The Bay Area required permits for units larger than 250 horsepower, and in Sacramento the threshold was 50 horsepower. In the smaller sizes, there was no requirement for permit licenses. As a consequence, only 12 of the region's estimated 5,000 units were subject to pollution controls.

A CEC report indicated that California should be able to force power plants that were allowed through expedited permits to sell their electricity only within the state. However, Bill Chamberlain, chief counsel to the CEC, told the commissioners many U.S. Supreme Court decisions had held that a state couldn't use its regulatory powers, including permitting processes, to give its residents an economic advantage. "I don't think these Supreme Court rulings would be modified by the fact that there is an urgent crisis." Neighboring states warned California that any attempt to confine power within California could be met with reprisals.

The CEC lawyers concluded that California could not require new power plant builders to dedicate their power to the State. This finding was despite Davis's pledge that any new power was to stay in California. The CEC was dubious that the emergency peaking plants installed in the state would be temporary. Peaking plants were, in theory, to operate only during peak periods. Many assumed the peaking plants would end up selling power to the DWR. The governor's office and the CEC expressed their opinion that the western states needed more power and that new plants would be needed.

The DWR threatened to refuse to buy power on the spot market if the price of energy became "ridiculous" but did not define ridiculous. Raymond Hart, DWR's deputy director in charge of power purchases, suggested that if prices became too high, the state might be better off ordering blackouts or implementing conservation programs designed to cut power use on short notice. Hart had to reconsider his stance because of the FERC's recent order that allowed generators to refuse to sell emergency power to non-creditworthy buyers. The DWR, however, was a creditworthy entity and mandated that emergency power must be supplied to the ISO. The ISO predicted that severe shortages could bring thirty-four days of rolling blackouts and warned that a refusal by the DWR to purchase would further aggravate the situation.

On April 9, PG&E petitioned the bankruptcy court to block parts of a recent PUC order. PG&E believed the PUC's accounting provisions had changed the rules retroactively and artificially extended the rate freeze. The PUC disagreed and characterized PG&E's petition as "the first step in the plan to deregulate itself." The PUC further argued, "PG&E may be able to claim an artificial end to the rate freeze, which could result in drastically higher electrical rates."

On April 9, U.S. district court judge Frank C. Damrell ordered Reliant Energy Services to sell emergency power to the ISO, but not before he blasted the FERC for its ruling that cash-strapped buyers of emergency power must be able to pay for the power. Judge Damrell said it was FERC's own words that he relied on in issuing the injunction pending the agency's decision on creditworthiness. He also canceled an April 12 hearing on a motion by the ISO to amend its lawsuit against the supplier in which it sought to convert the suit to a breach-of-contract litigation. He further stated any action in his court would be "futile." The appeals court, citing jurisdictional problems, eventually stayed his order. As the judge backed away from the fracas, he accused the FERC of dragging its feet and talking out of both sides of its mouth. Three of the state's lawmakers were to attend a FERC hearing in Boise, Idaho, focusing on the price volatility in the West and a request for a temporary price cap on wholesale rates. FERC said the ISO "misunderstood our order," which "required a creditworthy counter party, such as the California Department of Water Resources."

On April 23, the assembly passed SB 28x, a bill designed to expedite the construction of power plants. Jan Smutney-Jones said the measure "[would] make a difference in the time it takes to get a plant built. Trimming even a few months off the process has a significant impact in attracting new power plant development." The legislation was estimated to cost $3.2 million and would provide the following benefits: (1) Require cities, counties, and other local agencies to submit initial comments on power plant applications within forty-five days of its filing. Final comments would be due within one hundred days. Current law had no such deadlines. (2) Require the CEC to issue final decisions on certifying "re-powering" projects, the modernization of existing plants, within 180 days, instead of a year. (3) Extend for seventeen months, until Dec. 31, 2002, an expedited siting process for temporary power plants, known as peaking power plants. (4) Mandate that power plant owners require their facilities reduce air pollution and do so by January 1, 2004, under the auspices of the state Air Resources Board. (5) Require payment of unemployment benefits to workers in any week caused by unscheduled power outages resulting from shortages of supply. (6) Provide incentives for "distributed generation," in which homeowners or merchants could produce electricity for themselves and others using solar panels, wind turbines, or other such generating equipment. Utility fees, known as "standby charges," would be waived for up to ten years.

Calpine Corporation and Bechtel Enterprises proposed building a 600 Mw power plant in south San Jose's Coyote Valley. The plant, named Metcalf Energy Center, was to be built on fourteen acres, formerly a junkyard, next to PG&E's Metcalf substation, a site that would have been shielded from sight of the neighbors by a 350-foot hill. Cisco Systems, however, had planned to build a campus within the vicinity of the plant and objected to it. The San Jose mayor and city council supported Cisco's opposition; the governor recommended the CEC approve the plant and plant site; residents of the area took a NIMBY attitude toward the plant. Had the plant been approved, it could have been producing 600 Mw of electrical energy the following year.

In an intriguing aside, a new type of nuclear power source attracted the interest of some generating companies and government regulators. They had been following tests on new South African technology called "pebble bed modular reactor" or PBMR. The technology was still in the embryo stage, but it was advertised as cheaper, faster to build, and easier to operate than conventional light- water reactors that produced 20 percent of the nation's power. No new light-water reactors have been licensed in the U.S. since 1978, and there is still no long-range plan for storage of waste radioactive products. The PBMR is different from the traditional light-water reactor in several ways. A PBMR uses helium, not water, as a coolant. Helium cannot combine with others chemicals, so it cannot become radioactive when it passes through the reactor's core. Instead of using pellets of uranium oxide as fuel, the pebble bed uses tennis-ball-size spheres of silicon carbide-coated uranium oxide. The pebble bed has an inherent safety advantage in the fuel, as it never becomes hot enough to melt. As the fuel pellets circulate through the reactor, they generate only about one-tenth as much heat per square foot as a conventional reactor, partially because when stacked, the spherical fuel units provide a natural path through which the cooling gas can circulate. Each pebble bed reactor can produce 100 to 125 Mw. These smaller plants are faster to build and less expensive. Actual construction was estimated to take about two years. The pebble bed systems require shut-down for refueling every six years, as compared with the light-water reactors that are shut down every eighteen to twenty-four months. The pebble bed reactor produces one-ninth the nuclear waste of the water-cooled reactor.

But getting back to history, an assembly oversight committee was convened to determine whether gas companies had manipulated the natural gas market during the energy crisis. State regulators had complained to the FERC that El Paso and its affiliates made it oppressive for other gas companies to use the El Paso pipeline capacity to California. This permitted El Paso to reduce the natural gas supply, thereby increasing demand and driving up prices. PUC attorney Harvey Morris revealed to the legislature that El Paso had improperly sold 40 percent of its California pipeline capacity to its affiliate and had set unrealistically high prices, thereby discouraging use by others and driving up market prices. FERC dismissed this allegation. El Paso executives defended their companies, maintaining that weather, storage levels, extraordinary demand, and other factors beyond their control had affected those natural gas prices. Records showed prices more than quadrupled for the year ending in March. February prices at Topock jumped from about $2.59 per million BTU to $12.69 per million BTU, while national prices rose from about $2.61 per million BTU to $6 per million BTU. The PUC estimated that due to El Paso's market manipulation, the cost of natural gas in California would cost consumers more than $100 million annually. The Battle Group, a consulting firm, placed the cost much higher, estimating the cost for SCE alone at $750 million. A research firm, Lukens Consulting Group, concluded the high prices were the result of a complex chain reaction: "Higher prices caused an increase in demand for gas, which led to higher prices of gas, which increased demand for pipeline capacity, which caused an increase in the price of pipeline capacity." The PUC contended El Paso acted unethically by giving its affiliate an unfair advantage in bidding for pipeline capacity to California by structuring the auction to favor awarding the entire block to a single bidder. Twenty-four companies had submitted bids for capacity in the pipeline. Only El Paso Merchant bid for all of the capacity, offering $38.5 million for a fifteen-month contract. They also had failed to disclose to other bidders, except El Paso Merchant, that another affiliated company would soon offer discounted rates for use of an interconnection pipeline extending from Topock into Southern California.

El Paso scoffed at that allegation, arguing that the minimum price set for the pipeline capacity was a fair one. Auction terms stipulated that if the total of all the bids exceeded the "highest total package," the capacity would be parceled out. They further maintained that offering an interconnection lines discount to the bidders did not involve crucial lines. El Paso claimed all bidders had the same opportunity.

On April 24, Standard & Poor's downgraded California's credit rating from "AA" to "A-1 plus." This downgrade would increase California's borrowing costs and raised the question of where the State would purchase power in the coming summer. Bond ratings are the benchmark of a State's financial standing. The new rating status moved California's debt-payment from "very strong" to "still strong" but placed it in a class considered more susceptible to adverse effects due to changes in circumstances and economic conditions. The downgrade increased pressure on Davis to issue $10 billion to $15 billion in bonds in the summer as part of his rescue package.

On April 25, FERC adopted a compromise to limit California's power costs during emergencies. The plan would come into play whenever the electricity reserves fell below 7.5 percent, which was the ISO's Stage 1 emergency. California, however, would be required to file an application by June 1 to enter into a regional transmission organization with other western states or the new pricing plan would end. On a day-to-day basis, officials would determine how much it cost for the most expensive generator to produce electricity, and that would become the "proxy" price. It was estimated the price would range between $150 and $250 per mwh.

Williams applied for a certificate to add compressors and other equipment to its natural gas Kern River Transmission Company line. With prodding from the governors of California, Nevada, Utah, and Wyoming, the certificate was finally approved by the FERC. The project included new compression stations in Utah and California and upgraded compressors in Wyoming, Nevada, and Utah at an estimated construction cost of $81 million. The project would provide an additional 135 million cubic feet of natural gas for California.

The DWR set up new headquarters for the California Emergency Resources Scheduling, the group established to negotiate long-term contracts for electrical energy. The results of the negotiations, however, were closely guarded secrets. Davis released some partial figures; Davis and PG&E clashed over the interpretation of the figures. Davis blamed PG&E's bankruptcy for driving up prices. PG&E countered, blaming the State's increased spending as a result of the FERC ruling that forced the State to continue purchasing emergency power. No one had a complete answer, as Davis's office continued to hold the final information close to its vest. Meanwhile, the PUC had been waiting to receive the DWR's "revenue requirement."

Senator Nell Soto's (D-Pomona) bill SB 1x cleared the Senate Appropriations Committee on April 30. Through a "windfall profits" tax, it would allow recoupment of 100 percent of power charges that were deemed unreasonable. The proposal would have forced generators to pay the State of California all money collected above reasonable limits, to be determined by the PUC. Five companies were the principal targets of the bill: AES, Mirant, Duke, Dynergy, and Reliant. Any price above 8¢ per kwh would have been considered unreasonable and taxable as a windfall profit. Energy suppliers challenged the bill, arguing it would discourage energy producers from building any new power plants and/or upgrading existing plants in California and might even discourage them from producing power when the state needed it most. Republicans in the legislature opposed the bill.

Small generators, such as geothermal plants and gas fired cogeneration plants, were owed about $1.5 billion for energy they had supplied to the utilities. This only added to the debt the utilities incurred as a result of the retail rate freeze and the unregulated, out-of-sight wholesale prices. SCE had been sued for unpaid bills by about twenty-five small generators. Four generators sued PG&E and nineteen filed motions with the bankruptcy court seeking release from their long-term contracts, thereby allowing them to sell their energy on the lucrative open market. Two generators were allowed to shelve two contracts, five lost their fights, and five won the right to place a lien on SCE's assets. PG&E's position was that it could not afford to pay an extra $284 million per month if the small generators were released from their contracts and that the DWR would end up "picking up the tab." This additional cost would unavoidably revert back to the ratepayers.

On April 30, Governor Davis released his financial plan, declaring the State would spend more than $18 billion through June of 2002 for electrical energy. The plan indicated the state could survive the energy crisis without additional rate increases and without draining the State's budget. The planned projections were based on some trivial assumptions, including the ideas that the State would pay significantly less for electricity in the hottest summer months than it was currently paying, and that many of the QFs that had been refusing to supply energy because of lack of payment would start selling energy again.

In the middle of all this debate, the California Senate sent Davis an authorizing bill, SB 6x, to induce the State to get into the energy business. The bill allowed the State to establish a power authority and to build and run State-owned power plants. The public utility would have $45 billion available in bond money to build, operate, seize power facilities, and institute conservation programs.

Seeking help in the energy crisis, Davis summoned executives from power generating companies; ten attended in person and two others via telephone. He asked the companies to forgive 30 percent of what was owed to them by SCE and PG&E. One company immediately rejected Davis's proposal. Davis told the group that without the reduction, the legislature was unlikely to accept his plan to buy the utilities' transmission lines, but if his proposal went through, they could be paid this year. On the other hand, he was of the opinion that if it was rejected they would have to wait three to four years to get their money through the bankruptcy court. Senator Burton said "I'm going to insist they at least take [a] 30 percent haircut on the monies they're owed."

State legislators believed Davis's plan for saving SCE would be rejected if it were pushed to a vote without modifications, so they made an effort to design their own plans. Four separate plans were produced, one in the senate and three in the assembly. Davis's plan called for the State to pay SCE $2.76 billion for its portion of the transmission lines, to be financed through revenue bonds, coupled with a cash contribution of $400 million from SCE's parent company. SCE would be guaranteed a ten-year 11.6 percent profit. The legislature's four plans are described below.

Plan One: Option to Buy. Under this plan, the State would receive an irrevocable, five-year option to purchase the transmission lines for $1.2 billion. The money would be given only if the option were exercised. The State would dedicate a portion of the rates to pay off SCE's debts, which were estimated at $3.5 billion, but which would be reduced when generators wrote off 25 percent to 30 percent of what they were owed. This plan required a $400 million contribution from the parent company. The State would receive electricity from SCE's Kern County plant at a discounted rate. This option included possible joint ventures in future power plants.

Plan Two: Modified Grid Sale. The State would buy transmission lines for $2.76 billion, the same price previously agreed to by Davis. The deal was restructured so that no portion of the money would go to SCE's shareholders and more would go to retiring the debt. SCE's parent company would contribute $400 million for debt reduction. Generators owed money by the utility were to write off 25 percent to 30 percent of what they were due.

Plan Three: Corporate Acquisition. The State would acquire SCE assets if the company filed for bankruptcy. This option supported acquisition of PG&E as well, but allowed the SCE deal to stand-alone. This option also assumed assets of PG&E and SCE could be acquired at an extreme discount in bankruptcy court, perhaps $3.5 billion plus assumption of assets.

Plan Four: Utility Tax District. The State would not purchase transmission lines. A mechanism enabling utilities to refinance their debts with minimal State involvement would be formed. Special tax districts would be created and bonds sold.

Reliant's response was that any solution requiring a 25 percent to 30 percent reduction in debts owed to them was unacceptable. They felt key factors in developing an acceptable proposal would be contingent upon the cooperation of SCE, the power producers, consumer groups, and other interested parties.

An assembly Republican caucus believed the State should bite the bullet and pay for the electrical power used instead of borrowing money and delaying the fiscal anguish. Assemblyman Bill Leonard (R-San Bernardino) said the Democrats' plan would "saddle electric customers with higher utility rates to pay the State back with interest." State Treasurer Angelides denounced the Republicans' suggestion, arguing that the GOP's blocking of the bond deal would cut into the current year's spending for both education and transportation and would take the State into fiscal insolvency. It would also force the legislature to pass the bond bill with a simple majority. This would delay the bill's effective date for ninety days. The Republican minority criticized Angelides' proposed bridge loan of $4.1 billion as "expensive gimmickry."

The Republicans in the legislature refused to vote for the sale of $10 billion in revenue bonds until they received more information on the negotiated long-term contracts. They also questioned the numerous assumptions and estimates included in the plan. They were particularly uncomfortable with the assumption that the State would pay $195 per mwh on the spot market in the summer months. UCEI's Borenstein told the regulators, "There's a reasonable chance this summer [that we] will be paying $1 billion per week" for electricity. State Treasurer Angelides had serious concerns that if the bond bill didn't pass by May 8, his $4.1 billion bridge loan would be jeopardized.

The Republican proposal called for the State to sell $8 billion in bonds instead of the $12.5 billion proposed by Democrats. Their claim was that such a bond issue would relieve the pressure for electrical rate increases. They further proposed that any difference between the two bond plans could be made up with money from the general fund, money Governor Davis would take from various low-priority projects. Their plan also called for rebates to municipal utilities. Those who opposed such rebates regarded them as the taxpayers' money used to subsidize ratepayers whose utilities were in financial trouble. The Republican plan went further in reinstating the provision that would allow residential customers to make direct deals with electrical suppliers (direct access), and thereby would encourage power plants to be built quicker.

California Secretary of State Bill Jones proposed an alternative to Governor Davis's proposal. He thought the State should ask the power generator debtors to forgive some portion of the $14 billion they were owed in exchange for future immediate payment. In addition, he proposed that the State offer low-interest loans to the utilities to help them repay their debts, using an asset such as the transmission lines as collateral.

May

In early May, wholesale electrical energy prices declined dramatically, and California officials considered backing out of some long-term contracts unless suppliers agreed to lower their prices. However, the suppliers were particularly reluctant to comply. State officials had been riding high as a result of signing long-term contracts, along with the drop in wholesale prices. Private experts and analysts cautioned that backing out of the contracts would be a wrong step, especially since prices might climb again. The State had not released the details of the contracts, though they did report that the average price was $138 per mwh through the end of the year, and then it dropped to $70 for ten years. With the cooler weather the state had been currently experiencing, spot prices had dropped into the range of $50 to $60 per mwh.

The federal bankruptcy judge set a precedent on May 1 when he rejected PG&E's attempt to recover as much as $4 billion from the ratepayers by lifting the caps that were set to remain in place until March 2002. The judge said that the regulations allowed the State to exercise its "police powers" and that the State's authority to set "public policy" functions was not within the purview of a bankruptcy court. The judge ruled, "The fact that the result may be a negative impact on PG&E and a positive impact on PG&E's customers does not change the fact that PUC's rate-making implements public policy." The PUC was more than pleased with the ruling; if the court set rates, it would have negated the PUC's charter to set rates. PG&E said it "[would] continue to pursue all legal challenges"—in either federal district or appellate courts or through regulatory rulings—to lift the retail rate caps. PG&E maintained the PUC had changed the rules retroactively and artificially extended the rate freeze.

On May 1, Mirant received a license for a 530 Mw power plant to be built in the Bay Area but decided to delay construction of the plant due to the political climate in the state. Mirant spokesman Chuck Griffin said, "We're in a climate where things like windfall profits tax and the seizure of power plants [are being considered]. Because of the business climate we're faced with right now, we couldn't honestly say we're ready to go full bore." He denied Mirant was trying "to send some sort of a signal" by delaying the construction. He said Mirant was very much interested in proceeding with the plant. Gary Ackerman of WPTF agreed, saying, "The political risk of doing business in California has increased." Ackerman further claimed the political climate caused one power company to scrap a new power plant project and pull out of talks to sell long-term power.

The FERC had been pursuing a concept of regional transmission organizations, believing such a plan would better coordinate the nation's fragmented transmission grid. The FERC had approved a plan in April that involved imposing price caps on wholesale electricity sold in California whenever the state's electricity reserves fell below 7.5 percent. The program kicked in on May 5.

The state legislature, frustrated with FERC's failure to impose caps on wholesale energy sold in California, was expected to announce on May 7 that the legislature would file suit against FERC for neglecting its legal responsibility to ensure "just and reasonable" wholesale electrical rates. The tactic raised several legal questions about the legislature's ability to sue FERC to force it to act and impose "just and reasonable rates." FERC had recently adopted a compromise plan that had been widely criticized by California politicians. It was a one-year order that would impose price constraints whenever electrical reserves fell below 7.5 percent—a Stage 1 emergency. The benchmark price would be determined based on the highest cost of the least efficient power plant operating at the time of the shortages. Clark Kelso, professor at McGeorge School of Law in Sacramento, said, "There's just no way the legislature can show that it represents the legal interests of the people of California. The legislature can sue sometimes when a particular issue directly affects the legislature themselves. We are in a situation where FERC's decisions are having a substantial regulatory and financial impact in California." Bay Area attorney Joseph W. Cotchett agreed to take on the suit for $1 a month.

Democratic lawmakers, Senator Burton and Assembly Speaker Hertzberg, along with the City of Oakland, asked the Court of Appeals for the Ninth Circuit to impose caps on wholesale prices and provide refunds for previous purchases. The court rejected the filing, saying, "Petitioners have not demonstrated that this case warrants the intervention of this court." Throughout the whole crisis, the State had complained that FERC had shirked its duty under the Federal Power Act to set "just and reasonable" wholesale electricity prices.

Power use in the summer was expected to exceed 33,000 Mw, and in the late spring it was expected to be "touch and go," as approximately 30 percent of California's generating capacity was off line for maintenance in preparation for the summer. The ISO predicted potential blackouts when the demand exceeded 40,000 Mw, which had occurred thirty-four times the previous summer. An early May heat wave coupled with several power plants down for repairs and/or upgrades caused a series of blackouts in the late afternoon of May 7. The first was ordered at 4:45 p.m. Just three hours earlier; ISO director of grid operations Jim McIntosh had predicted in a press conference that the state would probably escape power interruptions for the day, so grid operators were surprised by the quick occurrences of the blackouts. At the time, Diablo Canyon had one unit down for refueling, the San Onofre nuclear plant had remained down for fire-related repairs, and many of the state's smaller plants were down for maintenance and/or air pollution upgrades. McIntosh subsequently reported that a combination of hot weather, four western nuclear plants down for refueling and maintenance, and a one-hour loss of imports from Canada had brought the California grid to within five minutes of blackouts. According to the DWR, this problem was exacerbated by the fact that BC Hydro had withheld promised power for several hours until it received money upfront from the State.

More blackouts were predicted as the heat wave continued into May 9 and exacerbated the situation. Several Central Valley communities were enduring record temperatures. California's neighboring states were also experiencing heat waves and power plant problems, thereby restricting their ability to export power to California. The DWR accused the power suppliers of deliberately withholding supplies in order to jack up prices but admitted they had no proof. The generators flatly denied that charge. The state also lost 1,000 Mw of power due to a transmission problem in western Canada, which resulted in blackouts that lasted from forty-five minutes to one hour.

The American Electric Reliability Council and the ISO made widely different predictions on how California's electrical climate would look in the coming summer. The ISO was predicting three hours of blackouts, as opposed to the Council's prediction of fifteen hours a week. The Council projected that blackouts would occur at various times during the day and not just in the late afternoon when the peak demand normally occurred. The Council predicted lower levels of power would be available from the state's power generation facilities, accompanied with lower imports of power from its neighbors. The ISO said the state could expect blackouts when the electrical demand exceeded 40,000 Mw, which as previously stated, happened thirty-four times the previous summer. The ISO estimated plant shutdowns would drop 2,500 Mw of power; the Council had estimated the number to be 4,500 Mw.

The federal government asked for bids to widen and improve Path 15 transmission facilities. The request for bids attracted a range of bidders, including East Bay Municipal District and power generators Mirant, Williams, and Calpine. Some independent experts warned that if a generator won the bid, it would be in a position to manipulate the electrical market. The ISO estimated the restriction of Path 15 added $221.7 million to the State's power bill in the sixteen-month period ending in December. UCEI's Borenstein cautioned that the Path 15 bottleneck created an incentive for power sellers to treat California as two separate markets, making it easier for the sellers to manipulate the market. Borenstein warned there were problems—if a generator was allowed to widen Path 15, it would be in a position to charge "tolls" to those companies who wished to use the lines. A power generator could conceivably withhold power, thereby increasing congestion on Path 15 instead of decreasing it, as was the intended goal.

In response to the high-energy prices, the energy industry had been making significant investments and efforts to build electrical power plants, drill for natural gas, and build pipelines. President Bush was prepared to announce a national energy program, while Vice President Dick Cheney itemized a list of deficiencies. This list pinpointed the shortage of power plants, oil refineries, natural gas resources, and gas pipelines, as well as other energy resources. In response to nationwide high prices, particularly in California, the electrical industry was expected to build 90,000 Mw of generating capacity over the next year and a half. This represented one-quarter of what the DOE had predicted would be required by the year

2020. Natural gas companies started building pipelines at a frenetic pace. FERC reported that 1,895 miles of new pipeline had been built in 2000, and it expected 4,300 miles to be built in the current year, and 4,650 miles in 2002. These would be unprecedented increases in capacity. Bush blamed high gasoline prices on environmental laws that hampered refineries from increasing capacity. Both Bush and Cheney attributed high-energy prices to manacling regulations, and both men wanted to streamline regulations. This course of action was applauded by industry. Jay Hakes, former director of the EIA, opined that the recent surge in energy industry activity indicated there was not a chronic supply deficiency. He said the problem was not that energy companies could not respond to demand, but that the newly deregulated energy companies were now denied the certainty of earning the return on their investments that state and federal regulations had once provided, and they tended not to respond too quickly. He went on to say, "Washington seems bereft of solutions to the real energy problem. Deregulation cut the fat [out] of the system and left people vulnerable to very volatile markets." He said the government needed to find a way to buffer the swings between surplus and scarcity. A possible role for governments would be to offer refiners and electrical producers tax incentives to build extra capacity to cushion against future supply shortages.

PUC president Lynch and PUC administrative judge Christine Walwyn wrestled with proposals on how to divide a $4.8 billion rate hike. Residential rates would increase an average of 11 percent to 17 percent and big industries' rate would jump 50 percent or more. Several commissioners wanted to select a new rate structure. In a special PUC meeting on May 13, new rates were approved, allowing SCE and PG&E to start collecting the higher rate on June 1. Walwyn recommended rates that were slightly lower for residential customers—an average of 11 percent hike for SCE and 15 percent for PG&E. Big industrial customers would pay an average 52 percent increase for SCE and 55 percent increase for PG&E. Households that stayed under the baseline rate would see no increase. Lynch began looking for ways to change power rates that were tied to wild wholesale rates at federal installations.

On May 17, President Bush released his national energy plan. The plan recommended that Congress and various government agencies use legislation, regulatory reforms, and executive orders to promote construction of more oil refineries and power plants and to increase production of nuclear and coal-fired energy. The Cheney energy task forces suggested paying for the initiatives through a combination of tax incentives, direct government funding, and market-driven investments by oil companies and other energy producers. Bush's plan stressed modernization of the U.S.'s transmissions systems by connecting separate grids into a single, nationwide structure. At the time, each state or region built transmission systems to meet its own needs without any thought of connecting them into a national system.

Cheney said the building of nuclear power plants, as well as extending the life of existing nuclear power plants, was the most environmentally satisfactory way to increase energy supplies. Utility executives welcomed the administration's support for nuclear power. They did, however, question the economic viability of constructing any new nuclear power plants in the near future. Environmentalists contested the claim that nuclear power was the cleanest source of energy, arguing that mining, enriching uranium, and the storage of the nuclear wastes were all environmental hazards.

Back on April 9, Vice President Dick Cheney had predicted that fossil fuels would remain America's primary energy source for "years down the road." He recommended increasing the supply of fossil fuels rather than limiting demand. He dismissed the notion that "we could simply conserve our way out" of an energy crisis. He argued that the solution was for the government to promote new domestic sources of oil and natural gas. He further stated there should be an all-out drive to build power plants, predicting the need to construct one new generating plant a week for twenty years. The administration proposed opening protected lands to oil and gas exploration, rolling back environmental rules that constrained the burning of coal, and constructing pipelines and refineries. Additionally, he estimated there was a need for 38,000 miles of natural gas pipeline construction. Cheney emphasized that drastic measures were required because the need was so great.

The Democrats proposed their own energy alternative. It called for temporary price controls, bans on energy development, and prohibitions against tapping the national oil reserves. They wanted to give consumers as much as $4,000 in tax credits for buying energy-efficient homes and cars, and to offer businesses tax incentives to invest in technologies and energy-efficient vehicles.

Governor Davis and fellow Democrats disparaged Bush's plan for not promoting more conservation, for emphasizing new sources of energy, and for not providing solutions to ease California's crisis. The utilities, on the other hand, welcomed the plan's recommendations to speed the building of new power plants, transmission lines, and natural gas pipelines. Keith Bailey, chairman of Williams, said, "It appears at first glance to be a very balanced document, with combinations of conservation as well as the need to create new . . . sources of energy." Most environmental groups opposed the plan.

The Bush administration's energy plan contained many features that would have allowed the energy industries to benefit financially, thereby encouraging more energy development. The plan emphasized increased development of fuel supply and transmission facilities necessary to transport energy throughout the country. The plan would promote easier access to public lands and expedite the review process for proposed refinery and power plant construction. The administration maintained that these steps, including a commitment to a diverse mix of fuels, would provide American consumers with an abundant supply and stable prices in the long run. U.S. energy generation by fuel type at that time was as follows:

Fuel Type	**Percent**
Coal	52%
Nuclear	20%
Natural gas	16%
Hydroelectric	7%
Oil	3%
Renewable energy	2%

Vice President Dick Cheney recommended the federal government be granted authority to assume rights-of-way over private property in order to expedite construction of additional transmission facilities. John Rowe, president of Exelon Corporation, was pleased the plan would ease license extensions for nuclear plants.

Congressman Joe Barton (R-Texas) introduced legislation before the House Energy and Commerce Committee to encourage large electricity consumers to reduce consumption and profit from reselling their power. The bill would also authorize the spending of $220 million to add a third transmission line to Path 15. Federal assistance, if requested by the governors, would be available to produce more power from federal power projects such as BPA. Environmental rules would be relaxed for new power plant construction or emergency power production. Democrats opposed the legislation, contending it would do little to ease California's power shortage during the coming summer and would provide waivers on clean air, thereby endangering species protection rules. Critics complained the legislation did not contain some sort of cost-based pricing of wholesale rates in order to prevent run-up in prices during shortages in the summer.

In a reversal of his environmental policy, Davis announced a plan to allow business and industry to use high-polluting diesel generators to produce electricity during anticipated blackouts. During Stage 3 emergencies, participating businesses could start their diesel generators while simultaneously disconnecting from their serving utility. The State would pay companies 35¢ per kwh for this power. Davis also offered the participating businesses 50 percent discounts on the air emissions credits they would need to comply with current smog rules.

A Field Poll released May 22 indicated Californians favored the building of more nuclear power plants (59 percent in favor), the capping of the wholesale price of electricity (70 percent in favor), and the relaxing of air quality standards, which would allow older power plants to be upgraded. The poll also showed Californians were not particularly enthused about the sale of $13.4 billion in bonds to pay for electricity along with the idea of using eminent domain to seize power plants. The opponents said the poll results could be misleading, as the nuclear issue had been dormant for so long. However, the majority of registered voters still opposed offshore drilling for natural gas and oil to ease the energy crunch.

Williams's CEO Keith Bailey suggested that FERC place temporary caps on profits for all energy suppliers between May 1, 2001, and the fall of 2002, when supply and demand were expected to become more balanced. His objective was to save California's public utility market and encourage the building of new power plants once power suppliers had the confidence that they would be paid for their product. Californians, in turn, would be assured that they were buying power at a reasonable price and that the suppliers would be receiving a reasonable profit. His suggestion was a wholesale rate equal to the cost of production plus a 15 percent profit, a figure that represented more than most suppliers normally sought. Williams had been owed $252 million for electricity sold to date, but Bailey was not insisting this debt be paid before his proposed cap took place. He recognized the outstanding debt would eventually be resolved either through the bankruptcy court or negotiations. Bailey was firmly opposed to any suggestion that the State of California seizes power plants to force delivery—he viewed it as a totally "draconian" step.

Duke Energy had purchased PG&E's Moss Landing Power Plant in 1998, and subsequently developed a project to add 1,060 Mw of gas fuel generation to the plant. Duke got four environmental groups to agree not to participate in any lawsuit or regulatory challenge that might slow or stop the project. In exchange, the groups would receive $1 million for monitoring and research of the plant's impact on the marine environment. Other environmental groups greeted the agreement with dismay. The environmentalists who were party to the agreement excused their action, arguing they had little chance of stopping the project, especially after approval by the CEC and the Central Coast Regional Water Quality Control Board. The $1 million agreement was part of a more than $12 million package of payments from Duke to civil, governmental, and environmental groups for monitoring of the plant modernization project. The $12 million mitigation payments were distributed as follows: $7 million to Elkhorn Slough Foundation to mitigate use of seawater; $425,000 to Monterey Bay Sanctuary Foundation over three years to monitor heated seawater discharge; $1 million to Monterey Bay Sanctuary Foundation spread over five years to monitor water quality via a program sponsored by Save Our Shores, Center for Marine Conservation, Friends of the Sea Otter, and Otter project; $100,000 to Marine Mammal Center to relocate its triage center for injured animals onto power plant property; $3.4 million to Moss Landing Chamber of Commerce over twenty years for infrastructure improvements in the Moss Landing community; $100,000 to Moss Landing for boardwalk to increase beach access; and $60,000 for an environmental assessment of a proposed Elkhorn Slough Circle Trail. If approved, Duke would provide an additional $250,000 endowment to maintain the trail.

June

California leaders were unhappy about the requirement that the state's three investor-owned utilities file proposals by June 1 detailing how they would participate in a regional transmission organization. Michael Kahn, chairman of the ISO governing board, said the ISO would meet the requirement but argued the requirement was not legal or appropriate. State officials were also concerned that the regional agency would deplete California's control, as the agency would then report to the FERC.

The ISO asked the FERC to revoke the rights of Dynergy, Duke, Mirant, and Reliant to charge free market prices. It argued the generators should be restricted to prices based on the cost of production plus a reasonable profit. In addition, the ISO had been asking for billions in refunds from the four generators. The ISO had made a similar request of Williams and AES the same week. Together, the six companies supplied 95 percent of California's power requirements. FERC already had in place a "mitigation" program in which wholesale prices were restricted when the reserves fell below 7 percent of the demand, at which time the ceiling price would be determined based on the production costs of the least efficient, most expensive power plant. The ISO belittled the mitigation program, saying the generators were able to get around the program through a ruse called "megawatt laundering," wherein power is moved out of state and then imported back into the state, thus evading the FERC's recent California ruling.

The DWR had staunchly refused to release the details of the thirty-eight long-term energy contracts they had negotiated with eighteen suppliers, believing that public knowledge of the details would hamper their negotiations. These contracts were worth $42.8 billion and allegedly contained confidentiality agreements involving 98,376,759 mwh of power to be supplied between June 2001 and July 2021. Assemblyman Tony Strickland (R-Orange County) filed suit, while several newspapers sought public disclosure of the DWR contracts. The Davis administration, faced with the possibility of losing the suit, announced on June 12 that it was going to release the details of those contracts. Calpine stated it had no problem with releasing the details in its three long-term contracts, which represented a total of nearly 2,500 Mw. Mirant California understood the need for disclosure but was still sensitive about its contract details, fearing some of the specific terms could be misused in the marketplace.

In a FERC hearing, Governor Davis demanded that the power suppliers should be required to refund California $8.9 billion of their electrical energy invoices. Two of FERC's commissioners were ready to back Davis's claims but said the energy companies should be given the opportunity to settle the issue within three weeks prior to the commission acting. Davis argued that the ISO had determined that the State was billed $8.9 billion more than what would have been warranted in an open competitive market for the thirteen-month period ending June 1. The commission had ordered reviews of about $124 million, while Davis, on the other hand, was seeking $8.9 billion in refunds. The FERC, in its order on June 18, requested that all parties sit down with an administrative judge to see if the refund problem could be solved. Of the $8.9 billion the ISO had identified as being in question, approximately two-thirds was debt incurred by the California utilities that was still owed to suppliers for past power purchases.

El Paso Natural Gas and its subsidiary El Paso Merchant came under FERC's scrutiny as a result of a twenty-three-page opinion reversing a March order which had stated that it wanted the administrative law judge to explore whether El Paso Gas and El Paso Merchant had conspired in exchanging secret information that gave El Paso Merchant the ability to obtain more than 1 billion cubic feet of pipeline capacity. The PUC, PG&E, and SCE had long contended that deal was responsible for the run-up of natural gas prices.

President Bush directed the FERC to invite "outside parties" to participate in the expansion of Path 15. PG&E owned the existing transmission lines that were capable of transmitting 3,000 Mw. Outside parties would be allowed to obtain the rights to transmit 800 Mw in the expansion. Any other utility or company wanting to transmit power over the expansion would be required to pay owners of those rights a fee and abide by their rules.

On June 27, San Diego superior court judge Linda Quinn ordered the DWR to make public the unedited versions of their thirty-eight long-term contracts. Earlier in the month she had agreed with Governor Davis's administration that certain details of the contracts could be withheld. The State was given until July 5 to act.

Many economists, including those who shy away from any government involvement, supported California's demands for a cap on wholesale power. California officials argued that the FERC mitigation plan didn't go far enough, as it allowed energy producers to get around the plan through megawatt laundering. UCEI's Borenstein opined that because the cap would fluctuate, sellers would have an incentive to withhold power to get the cap raised. Power generators and some economists said the price caps would encourage suppliers to ship power out of state and would discourage the building of new power plants that were desperately needed in California. Economists said the key to setting price caps was finding the right level; if the price caps were set too low, power supplies would disappear. The cap must instead be set high enough so suppliers would be encouraged to continue to supply power.

July

To recap the situation at the time: the DWR was purchasing power for the utilities through long-term power contracts; the California ISO was running the grid, using auctions to buy peaking power, which in theory was the optimum way to get the cheapest power; and the ISO was using some of the auctions to relieve congestion in transmission lines and using others to stabilize the grid. The result was that the power purchased was not necessarily from the cheapest source. For example, at the instructions of the ISO, SCE reduced cheaper power out of the Mojave plant by 5,660 mwh between May and July. The ISO then replaced it with more expensive power. The DWR had not participated in the ISO auctions, even though the ISO had requested them to. However, the DWR lost $14 million by selling excess power for less than it paid for it (the result of take-or-pay long-term contracts).

The ISO had been relying on the older peaking power plants spread throughout California to meet peak requirements for a few hours each day. These were the least efficient units, with heat rates in excess of 16,000 BTU per kwh; modern peaking units could obtain heat rates of around 7,000 BTU per kwh. Despite their inefficiencies, these older units were used to determine the wholesale price of electricity under the recent ruling by FERC, which said the price would be capped at 85 percent of the highest cost, least efficient plant operating at the time of a Stage 1 emergency.

For California to remain comfortable with power production, it had to maintain a 7 percent reserve, according to the guidelines of the time. Failure to meet that reserve level could activate the three stages of alert. The power needs of all the western states had continued to grow, with California's and Oregon's demands growing the most. There were now 32 million customers in California and the state's electrical demand had increased 14 percent. At the same time the supply decreased by 3 percent, resulting from the shutdown of older power plants, as well as a decrease in supply from other existing plants. It was estimated that on the hottest days, the state demand could be 50,000 Mw, including the 7 percent reserve. However, the ISO had not experienced a demand exceeding 41,000 Mw, in spite of the hot summer days. Contributing to the supply problem was the fact that neighboring states were facing increased demand also, and those states were not fond of California's aggressive power demands. On the positive side, however, Californians had stepped up to the plate and exercised conservation practices.

The cost of energy began to recede as the Bush administration put together its national energy plan. Bush, Cheney, and other officials credited the energy price spikes to under-investment linked to regulatory impediments, but analysts argued the crisis may have been the result of business decisions by energy companies. The record low prices of the 1990s, coming at a time of robust economic growth, had not encouraged investment in new oil and natural gas sources or in electrical generation. In 2000 the number of operating, drilling natural gas rigs totaled 550. With higher gas prices, energy companies responded with increased drilling activity. As of July 2001, there were 1,050 rigs drilling for natural gas, without any new lands being opened for exploration.

On July 2, State Controller Connell released details of forty-one long-term contracts, twenty-seven consulting contracts, and five short-term deals, along with stacks of receipts and invoices. She planned to subsequently release information on how much the State had paid individual suppliers on the spot market. The contract details included so many intricate pricing formulas that she was unable to predict exactly how high the hourly cost could have gone. Her chief deputy controller of finance said the price would be well in excess of Davis's predicted price of $69 per mwh; contract prices ranged from $55 to $249 per mwh. The DWR subsequently revised the forecasted price of $69 per mwh to $70 per mwh through the end of 2010, the higher price coming first. Some of those stated prices could decline if there was a decline in the price of natural gas. The long-term contracts had been released June 15, with power plant names, performance data, delivery points, and other details blacked out. In one power swap, the State agreed to send Powerex, the trading arm of British Columbia's government utility, 2.5 times more power in the spring than Powerex would return in the summer. As a result of the power purchase contracts, it was evident Californians would be paying premium prices for years to come in order to cover the cost of new power plants coming on line. A March 29 contract with Calpine resulted in the State paying Calpine $232 per mwh for 300 Mw from July 1 through September 30. A more recent contract with Calpine had been for about $469 per mwh.

With temperatures soaring on July 3, wholesale power prices in New Mexico and Nevada were reported as high as $135 per mwh, which was well above the FERC ceiling price of $91.87 per mwh. The price in Southern California went to $106 per mwh, while the cap for the State was $95. Many speculated FERC's price caps caused suppliers to limit power supplies. Power generators complained about FERC's new pricing system, saying they were confused about how much they could charge. They argued the price caps limited profits and discouraged sales to California.

On July 5, the California attorney general requested that the SEC examine the financial transfers that PG&E had made to its parent, PG&E Corporation. Between 1997 and 1999, $4 billion had been transferred from PG&E Company to PG&E Corporation. The attorney general wanted a meaningful review by the SEC to determine what impact this transfer of assets may have had on the bankrupt utility. Greg Pruett, vice president of PG&E Corporation, maintained, "That is an issue that has been investigated multiple times by the Public Utilities Commission earlier this year and by an assembly oversight committee, and each of these audits indicated there had been no conclusion that the transfers were inappropriate or that they were injurious to PG&E Company." Pruett said the bulk of that money paid off debt incurred in building power plants and the balance went to pay shareholder dividends.

On July 6, PG&E announced an agreement with Calpine for 450 Mw for five years at a price of 5.37¢ per kwh. The agreement was submitted to the bankruptcy judge for approval. The price was the same that had been recently set by the PUC for long-term contracts and was higher than what PG&E had been paying for energy lately, but it was less than one-third the price paid in January under a formula that factored in the cost of natural gas.

On July 9, the DWR released 1,770 pages of invoices and trade confirmations that provided a very detailed look at the State's energy purchases from the time it started buying power on January 17, 2001 all the way to the end of May 2001. The State had made the following purchases:

Company	Amount of Purchase
Sempra	$429 million
PG&E Energy Trading	$23.7 million
LADW&P	$331 million
BC Hydro	$1.05 billion
Mirant	$1.24 billion
BPA	$167 million
SMUD	$80.7 million

On July 9, Curtis L. Wagner, chief judge of the FERC, was ready to announce his assessment of what California might be owed in energy overcharges. California had demanded $8.9 billion, which Wagner said was too high, suggesting the amount might actually be closer to $2.5 billion. The power suppliers had offered $716 million in refunds. Of that total, Reliant, Mirant, Duke, Williams, and Dynergy offered $510 million, BC Hydro offered $125 million, and California municipal utilities made up the difference of $16.5 million. PG&E and SCE wanted to pursue talks further, indicating they were willing to talk to anybody at anytime to seek a settlement. Wagner commended Davis for indicating that he might not require a cash settlement but would consider other options, such as a renegotiation of long-term contracts. The FERC negotiations failed to produce a settlement by the deadline, and Judge Wagner said there would be no extensions. The five FERC commissioners were to begin fact-finding hearings on how much was truly owed to California from overcharges. Wagner said his recommendations were to go back no further than October 2, 2000.

On July 25, the FERC ordered further hearings to determine the amount that each of the involved electrical generators would be required to refund to California for alleged overcharges for electricity purchased between May and October of 2000. The sales amounted to about $2 billion for electricity that DWR purchased from suppliers in that year; the State had been asking for a refund of $3 billion. Only DWR's purchases through the defunct PX or through the ISO would be considered for refunds. Davis filed a formal request for a rehearing on the issue.

The DWR announced that current electrical rates were sufficient to pay the utilities' expenses and cover the State's purchases. Joseph Fichera, a Wall Street Financier who was advising Davis, believed the updated projections to be accurate in portraying future costs. In his view, the DWR needed slightly more than half of the billions of dollars, which would be generated by a March rate increase. The balance of the money from the increase would have been enough to cover the utilities' internal costs. Fichera also believed there would be sufficient money left over to pay the billions of debt owed to SCE's creditors, though at the same time he recognized the creditors would be unwilling to wait for payment. He recommended that the State pass a rescue plan to keep SCE out of bankruptcy danger. The senate had passed a rescue package, but SCE said that package did not go far enough.

Davis's agreement with SCE had an August 15 deadline, which would occur toward the end of the legislature's summer recess. Davis, anxious to get the agreement approved, threatened to call a special session to keep legislators in Sacramento if they did not approve the SCE plan before their summer break. Senate Democrats had been reviewing a plan by Senator Byron Sher (D-Palo Alto) that would have required large businesses to absorb the biggest share of SCE's debt and further required the suppliers to reduce their SCE debt demand by 30 percent. This would leave about $1.2 billion in debt that SCE would have to recover from the ratepayers. The plan would include the option for the State to purchase SCE's transmission facilities at book value, less than half the price Davis had offered to pay. Members of the assembly were, at the same time, drafting their own version of a plan. Their version would also separate users into large and small entities. Under their plan, most residential customers and small businesses power would come from the cheapest power sources—SCE's own plants and the QFs. Larger users' power would come from the long-term contracts entered into with the DWR.

August

On August 1, PG&E filed a complaint protesting the PUC's proposed "rate agreement" with the DWR. The agreement essentially said the PUC would agree to raise rates whenever the DWR needed more money to buy power. PG&E called it a massive "blank check" for the DWR and urged the PUC to hold public hearings to justify costs on any rate increases. The rate agreement was crucial to the State's sale of $13.4 billion bonds to pay past and future power purchases.

A congressional hearing on August 2 questioned the ISO's role in the sale of power to the DWR. Congressman Doug Ose (R-CA), who chaired the hearing, charged that the close relationship between the ISO and the DWR could result in the State forfeiting any claim to refunds for overcharges by the power generators. The issue was a five-member board appointed by the governor to run the ISO. The ISO, however, under federal law, was supposed to be independent of the State. The DWR, the ISO's biggest customer, was using State money to deliver power to the cash-starved utilities. Earlier in the week, the FERC had determined the DWR, because of its special access to information, had a competitive edge over other purchasers and should not be entitled to any rate refunds on its power contracts. Terry M. Winter, ISO president and CEO, argued that in January, when the DWR started buying power, there were no other buyers willing to buy power. "That occurred because the market was not functioning; they were the only creditworthy purchaser purchasing power." Kevin Madden, the regulatory commission's top lawyer, commented that the relationship between the ISO and the DWR raised troubling questions, and DWR was expected to agree that the ISO should be reorganized to assure its independence.

An ISO spokesperson acknowledged that an incident on August 9 brought the balance between electrical supply and demand so precariously close that the loss of one generator could have triggered the western region's worst outage in five years. Unfortunately, these incidents were occurring with alarming frequency. The assistant director of WSCC was quoted as saying, "We have seen a number of generators not responding to our orders to dispatch, and in this case we did not get the power that was scheduled through our grid." He went on to say, "It's very important in the operation of the electric system [to] abide by the reliability requirements." In one case a generator was scheduled to provide more power than it was able to supply and fell short by 1,600 Mw. Another generator did not provide the power the ISO had ordered, and the ISO asked federal regulators to investigate this incident.

The Bush administration had called for the building of one power plant a week for the next twenty years. A few months later, industry analysts, in contrast, warned of an electrical supply surplus. These mixed signals caused some power generators to rethink the wisdom of constructing more power plants. However, this was a short-term mentality; long-term forecasts may have proved to be correct. Some contradicted the fact that there was a crisis at all.

For example, Raymond C. Niles, a natural gas analyst at Salomon Smith Barney, said, "Power is cyclic now, like oil and natural gas. It's going to be moving like a pendulum now—tight supplies and abundant supplies, high and low prices." Christopher R. Ellinhaus, an analyst at Williams Capital Group, said power suppliers had proposed building 350,000 Mw of capacity by 2004, but with the falling power prices, it appeared only 100,000 Mw would be built. Christian Poindexter, chief executive of Constellation Energy Group, said if all of the proposed plants were built, the country would have half again as big a reserve cushion as energy planners considered necessary. Jone-Lin Wang, director of North American power at Cambridge Energy Research Associates, opined that in addition to infrastructures issues, there were signs of overbuilding in a number of regions where power plant construction over the last two years had been prolific, saying developers had a herd mentality. A spokesman for Cogentrix said a 1,100 Mw plant was scrapped in West Virginia after the company concluded rates were unlikely to justify the investment. Calpine had projected to bring 70,000 Mw on line by 2005 but later said they were not locked into that number. Barry Abramson, an analyst at UBS Warburg said, "There are some power plants being delayed or canceled due to economic conditions, not federal power caps. Projecting ahead, the companies are seeing increased supply in the area and lower demand than they thought." According to RDI Consulting, which advises energy companies nationally, only about three-fourths of the new generating capacity that had been expected to be available in the summer really came on line. "We expected 46,000 Mw of new generation this summer and got 35,000." Analysts predicted regulators would have a rare chance to determine how to create market conditions that would smooth out the power industry's boom-or-bust cycle.

FPL Energy wanted to build a 560 Mw natural gas fueled power plant on ninety acres in Rio Linda. This property had previously been zoned for industrial use to build a plant that would convert rice straw into ethanol, but it had never materialized. The CEC was expected to license the power project the following year, with the new plant going on line in 2004. The residents of Rio Linda and Elverta objected to the plant because they felt left out by local government. The arguments against the project included air and noise pollution, loss of ground water supplies, and negative impact on property values. However, they failed to realize that the plant, in addition to producing power, would have employed twenty to twenty-five employees, generated $3 million in property taxes, and $37,500 in sales taxes for Sacramento County.

After a hearing at which Governor Davis vilified Duke Energy, even though it had not been allowed to make any presentation to show its case, the out-of-state energy producers started fighting back. Duke's representatives had been kept out of the hearing room, while at the same time legislators at the hearing heard from three former SDG&E employees, who accused Duke of avoiding supplying full-capacity power from its plants. The governor and senate committee didn't hesitate to publicize the testimony of the three employees who had directly accused Duke of restricting output from one of its power plants. Duke representatives displayed excerpts in newspaper articles showing Duke actually charged less for power than its competitors. Mirant bought half-page ads in the *Sacramento Bee* and other major Northern California newspapers saying "It's going to be a long hot summer in California . . . but Mirant is going to help out."

State Republican legislative leaders, observing the magnitude of the DWR's energy purchases, opined that public investigation was in order and called for oversight hearings. In a letter to both the senate president pro tem and the assembly Speaker, Republicans asked for "comprehensive legislative hearings regarding all long-term contracts that had been entered into." The hearings were to include details regarding spot market purchases that had been made by the Davis administration. The lawmakers said the hearings should also address potential conflicts of interest by energy traders who had bought power and negotiated contracts. The Republicans went on to say, "Republicans and Democrats in the legislature have given the governor tremendous leeway to solve this problem. Now it is time to take a dispassionate look back and see what we can learn. We want [the] sun to shine in the process." Davis's spokesperson said, "It was simply more politics. I'm trying to figure out how it helps California weather the energy crisis by doing Monday morning quarterbacking."

California regulators discovered an email message that had been sent by an El Paso Corporation individual to a colleague regarding El Paso's pipeline capacity into California, a message, the regulators contended, that contributed to the sharp increase in California's gas and electricity prices. The message, from a vice president of El Paso Merchant Energy, said,

"Mojave is willing to do something along this offer," followed by a description of the offer and, "I will not officially request this discount until next Wednesday." A year and half later, a FERC administrative federal judge concluded the PUC had a strong case of market abuse, saying, "The burden of proof now shifts to El Paso." The PUC filed a complaint with the FERC asking for refunds of at least $184 million, arguing the damage had been considerably more. SCE claimed El Paso overcharged them by an estimated $864 million. An SCE attorney said that over a year's time, El Paso was chiefly responsible for raising California's natural gas cost by $1 billion and electrical costs by $2.7 billion—a total additional cost to California of $3.7 billion. El Paso argued it had done nothing wrong and blamed the surge on demand and not market price increase manipulation. At issue was the fact that El Paso owned the gas pipeline from Texas into California known as the "swing pipeline." On February 7, 2000, El Paso's subsidiary El Paso Natural Gas had initiated a weeklong open auction for rights to one-third of the pipeline's transmission capacity. Approximately twenty-five bidders participated in the auction, but in a quirky move, the contract was awarded to El Paso Merchant Energy. The PUC accused the bidding of being rigged. Additionally, the PUC said the move was made viable by a shipping discount Merchant Energy had obtained from a third El Paso company, Mojave Pipeline. Mojave Pipeline controlled alternate pipelines in California. The PUC argument was if all bidders had the same discount, the bidding process would have been more competitive. A member of the SCE consulting firm Battle Group said that with so much control over the pipeline, Merchant Energy had the capability to withhold capacity, thereby driving up prices. SoCal Gas, according to SCE, was put in a position that it couldn't wait any longer to purchase natural gas for storage, thereby putting additional demands on an already constrained system. Norma Dunn, El Paso vice president, said, "All of the regulations and rules that FERC set out for affiliate transactions were followed to the letter." An SCE consultant told the FERC that when the El Paso pipeline contract became effective in March 2000, the price of natural gas was $2.84 per million BTUs; by December the price was an average of $25, peaking at $59 on December 12. El Paso critics noted that when the contract expired fifteen months after it began, the price of gas fell.

The PUC drafted a decision providing that PG&E would pay more for the power it purchased from the DWR than would any other buyer of DWR power. The rationale for this decision was that the State was able to purchase cheaper power for SCE as more power plants could be accessed and that there was less transmission line congestion. In a press release, PG&E stated, "This massive cost shift proposed by [the] DWR discriminates against PG&E's customers and has not been subjected to public review, due process, or cost justification. This would lock our customers into 40 – 50 percent higher rates for DWR power over the next ten years, [as] compared to customers of SCE and SDG&E." The PUC draft called for PG&E to pay the DWR 1.4¢ per kwh compared to SCE's 1¢ and SDG&E's .9¢ In addition to the rate proposal, there were six other proposals, all of which were an effort to clear the way for the State to issue its $12.5 billion in bonds for the DWR's payment for electricity. The additional proposals included denying DWR permission to bill utility customers $600 million for various electricity conservation programs, as such billing had not been authorized by the legislature; imposing a 12 percent average rate increase for customers of SDG&E; and shutting down, effective retroactively, a program allowing customers direct access. This action was to ensure that all ratepayers would buy DWR power.

The lawmakers came up with a plan to rescue SCE from bankruptcy, and key members of an assembly committee wanted SCE to guarantee that the funding of the $2.9 billion plan would keep SCE out of the bankruptcy court. Bob Foster, SCE vice president, said he was confident that SCE could regain financial solvency and repay its debts without any risk to the State. He told the committee, "Can I give you a guarantee? No, I can't." Without such a guarantee, lawmakers remained unconvinced regarding SCE's ability to avoid bankruptcy. The proposed plan would permit SCE to sell $2.9 billion in bonds to be repaid by their ratepayers and included a five-year option enabling the State to purchase SCE's portion of the state's transmission system. Terms of the transmission sale were troubling for Foster, as he said such a sale to the State would impair SCE's borrowing capacity and reduce its prospects for becoming solvent. Additionally, SCE would want the fair market value, not the book value, for the 12,580-mile transmission system. Many members in the legislature from both parties had misgivings regarding the plan as it was presented.

On August 13, Davis was scheduled to appoint four members to California's new power authority. This authority was given $5 billion of bonding approval in order to finance generating projects that were to guarantee cost-based rates as well as promoting energy-efficiency programs. One billion dollars of that bonding was to be dedicated to loan programs for conservation and renewable energy projects. Environmental groups in opposition believed more money should be dedicated to renewable energy to reduce the use of natural gas and avoid its volatile pricing.

Consumer advocates were concerned about the PUC's preparations to adopt a new policy. PUC president Lynch released details of the new policy on August 23, titled "The Promise." This new policy was designed to protect power needs and supply sources for the foreseeable future.

In a meeting on August 24, the California Consumer Power and Conservation Financing Authority (California Power Authority) was inaugurated. Board members provided an outline of objectives, including plans to create a strategic reserve of publicly owned or controlled electricity. They targeted as much as 3,000 Mw of reserve by the summer of 2002. The board approved $1.5 million in startup costs and authorized seeking a top financial officer. The board wanted the agenda for the next meeting to include a briefing on conflicts of interest. While the first meeting provided very few specifics, State Treasurer and board member Phil Angelides commented that the authority should be "strong enough and entrepreneurial enough that it will be a competitor to the private sector." Skeptics admonished that the State might be headed down the wrong path if it tried to compete with companies that were more capable of operating an electrical business. Jan Smutney-Jones said, "If the power authority is designed to complement and supplement what's going on in the private sector, fine. If they're trying to compete, ultimately the taxpayers will lose a lot of money." Secretary of State Jones called it "the birth of yet another useless bureaucracy."

In response to questions from Republicans, the FERC confirmed that the State's new power authority would be required to get federal permission to enter into power contracts, including the building of additional hydroelectric plants or transmission lines.

August 27, an assembly committee voted support for a five-year option to purchase SCE's transmission lines for $2.4 billion, which was double the book price. Originally the State's rescue plan had called for the purchase at the book price. Additionally, the agreement included an option for the State to purchase SCE's transmission facilities with the legislature's approval. It included public conservation easements to approximately 24,000 acres of sensitive lands near SCE's hydroelectric plants and provisions allowing businesses, under certain conditions, to contract directly with generators for power if the state was faced with a supply shortage. Finally, it would require SCE to increase its supply of renewable energy, which Senator Burton called a "nonstarter."

The Democratic assembly's point man Fred Keeley opined that bankers had warned that the 3,600 large businesses in the SCE service area were inadequate to support a $2.9 billion bond issue. Supporters contended the transmission lines would be a valuable asset and would give the State more influence over the maintenance of the transmission system.

Opponents in turn argued that the State would commit to a costly maintenance and improvement program and was taking on an operation that was better handled by private business. An SCE vice president said the State's option to acquire the transmission system at book value would impair SCE's borrowing capacity and its creditworthiness. An angered Doug Keller of the FTCR called the option a bailout for SCE and promised to lead a voters' revolt. Supporters argued that if SCE went into bankruptcy, California's energy future would be jeopardized, sending the wrong signal to businesses and making it harder for the State to remove itself from the electrical energy purchases.

In August of 2001, energy use dropped by 6.5 percent. A CEC director said there were only four days in August on which the electrical demand exceeded 40,000 Mw, the amount believed to stretch the system to the breaking point. Energy officials credited higher electrical rates along with the financial incentives of the 20/20 conservation plan for the reduced usage.

September

State Treasurer Angelides, eager to sell $12.4 billion in bonds to cover the DWR's purchases of electricity as well as to relieve the drain on the general fund, was frustrated by the state regulators' foot dragging in passing a package of measures he believed necessary to satisfy investors. He released a statement on September 6 that said, "Until the Public Utilities Commission takes the actions required to implement the Department of Water Resources energy bond program, the State of California cannot move forward to market and sell the bonds." He went on to say he would release a new schedule on the bond issue only after the PUC had acted on the measures. But the package that the legislature, DWR, and Angelides had urged the PUC to approve caused the utilities to bristle. The utilities attacked, saying state law gave the PUC no authority to review State spending to ensure its reasonableness. They wanted the DWR to explain its spending in more detail and were very dissatisfied with how the wholesale power rate would be divided. At the same time, PG&E's customers had been deluging state regulators with hundreds of phone calls and emails displaying their anger at having to pay more for their electricity than did SCE's and SDG&E's customers. The PUC's rationale for the division of the costs was that if it was divided equally among the utilities, it would have been unfair to SCE's and SDG&E's customers and would be asking them to subsidize the more costly power going to PG&E.

The SCE rescue plan, SB 78xx, was approved on September 8 by a narrow vote in the assembly and was sent to the senate for action. The senate's passage was questionable, as Senator Burton openly opposed it. The bill, if approved, would have authorized SCE to sell $2.9 billion in bonds to be repaid by business ratepayers; the additional $1 billion necessary to clear SCE's $3.9 billion debt would be SCE's responsibility.

Governor Davis, invoking his constitutionally vested powers, seized the PX's January and February energy contracts. These were contracts that the PX had wanted to sell to recoup the millions of dollars the two utilities owed it for previous power purchases. Duke sued the governor, alleging Davis took control of long-term contracts that required Duke to deliver energy at a generally cheaper price than the going market rate. The U.S. Court of Appeals for the Ninth Circuit agreed with Duke, ruling two-to-one that Davis lacked the authority to take the contracts because Congress had not given the states such rights in energy matters. Duke also argued it should have been able to sell those contracts to other utilities for a greater profit than they were originally sold to the PX for.

Senator Byron Sher (D-Palo Alto) introduced SB 532, which would require every power producer in California to provide increasing amounts of power through renewable energy, the target being 20 percent by 2010. Many states had adopted "portfolio standards" requiring utilities and energy producers to allocate a share of their production to be derived from renewable energy sources. The amounts ranged from 2 percent to 30 percent, depending what was defined as a renewable source. Free-market advocates and some utilities were concerned that these requirements might drive up the price of electricity by rewarding inefficiencies in energy sources that were otherwise too expensive to stand on their own. These critics believe the state should consider potentially low-cost power such as coal, large hydroelectric, and nuclear power plants. Natural gas plants generated over half of California's power, and that figure could rise to 65 percent by 2010 if all currently proposed plants were built. The Wind Energy Association said wind turbines were getting larger and more dependable, and with the aid of federal tax credits they could produce power between 3¢ and 6¢ per kwh depending on location. With wind energy costs at the low end, wind energy became competitive with natural gas. Geothermal at the time ran between 4.5¢ and 7¢ per kwh, though it could have gone lower if a new round of federal tax credits supported it. In 2000, the CEC estimated the breakdown of electrical energy in California to be as follows:

Type of Energy	Percent
Natural gas	35%
Large Hydroelectric	19%
Nuclear	17%
Coal	16%
Eligible renewables	13%
Geothermal	5%
Small hydroelectric	3%
Biomass	2%
Wind	2%
Solar	<1%
Other	1%

The rate increases magnified unemployment in the state. This was due to the closing of industrial plants that couldn't meet payment of their energy bills. Shasta Paper Company laid off 400 workers on August 20 and closed its plant after its electrical bill jumped from $800,000 to $1.3 million. The plant manager believed the closing was just the "tip of the iceberg." Economists at UCLA remarked that energy woes could push California into a recession. The California Manufacturers and Technology Association recorded sixty-six energy shocks to the state, including layoffs, shutdowns, and business opportunities lost to other states. Many economists believed California could weather the increased rates, rationalizing that Californians used less electrical energy than the rest of the nation. The energy problem may have been less of a threat than the slowing economy, but it made its contribution.

Direct access, the freedom of customers to buy electricity directly from generators, was on schedule to be terminated this month, but it could have been revived by the legislature. The objective was to assure that all purchases of electrical energy would be made with the DWR. With the DWR purchasing electrical energy and entering into long-term contracts, the retail customer would be shielded from the high cost of energy. A consulting firm, Cambridge Energy Research Associates, teamed up with UCLA to create an energy forecast that indicated California unemployment would increase the current year by .5 percent from 4.9 percent, followed by another 1.1 percent the following year. California's average electricity cost, the highest in the continental U.S., went up from 1.05¢ to 1.2¢ per kwh, according to federal statistics. California's heavy-duty manufacturers and industrial firms were the biggest users of electrical energy and carried the brunt of the rate increases, with bills going up 50 percent or more.

On September 20, the PUC voted three to two to terminate the direct access provision of deregulation, effective immediately. The court's decision was one of the four steps the state treasurer had requested to enable him to move forward with his $12.5 billion bond issue. In addition, the PUC also raised retail rates for SDG&E customers by 1.6¢ per kwh. The PUC delayed action on two other measures the state treasurer said were needed in advance of the sale of the energy bonds. The first was a decision on how much each utility should be required to pay the State for electrical purchases. The second was an implementation of the law that stated the PUC had no right to examine whether State costs were fair.

After lengthy deliberation, the DWR filed a $179.5 million suit against PG&E in the U.S. bankruptcy court. The court had advised the State that if it wanted to collect, it would have to enter the bankruptcy case. The action would put California at risk to potentially lose its right to avoid being sued over its policies and practices. Steven Felderstein, in filing the DWR's claim, argued that the prevailing rule in federal courts was that the State's immunity was waived for this claim only.

On September 20, PG&E proposed a revolutionary restructuring plan to the U.S. bankruptcy judge. The plan as submitted asked the court to override state laws and regulations governing PG&E's ownership of its power plant hydroelectric and nuclear facilities. The plan was a complicated mixture of debt refinancing and corporate spin-offs that would, as it was represented, restore both PG&E's soundness and creditworthiness. It further stated that no additional rate hikes or a State bailout would be needed. The proposed plan would separate the utility from its parent company and pay off PG&E's $13.2 billion accumulated debt. PG&E's proposed reorganization plan would include the following: PG&E wanted to (1) split the parent company, PG&E Corporation and a scaled-down utility into two unrelated companies; (2) override state law in order to transfer State-regulated power plants to the unregulated PG&E Corporation; (3) sell power from these plants to the regulated utility for twelve years at prices well over the permitted rates, after which they would have the ability to charge market rates; and (4) give shareholders of PG&E Corporation stock in the new utility company. For its part, PG&E pledged to repay all its debts—approximately $13.5 billion—retain all employees, get out of bankruptcy by December 2002, and avoid rate increases.

The transfer of PG&E's hydro and nuclear facilities to unregulated status would violate state law that specifically prohibited PG&E from selling any of its generating facilities before 2006. To accomplish the proposed restructuring, PG&E was asking the bankruptcy judge to override state law, but there remained the question of whether he could do so. The plan would have involved a series of property transfers and bond offerings splitting PG&E Corporation into two separate, publicly traded companies. The PG&E utility company would sell retail power and would sell other assets to the second company, which would include its hydro and nuclear generating plants, its transmission facilities, and its natural gas pipelines. The sale would amount to $4.06 billion in cash and IOUs, and the shareholders would receive stock in the scaled down utility, details of which still had to be worked out.

PG&E's proposed plan to restore itself into a viable functioning utility encouraged SCE's creditors to consider the possibility of involuntary bankruptcy for SCE. Davis and SCE officials, however, continued to promote the State-sponsored rescue plan. Members of the legislature still were not entirely convinced the State plan was the appropriate solution. Debra Bowen, chair of the Senate Energy Committee said, "I hope [the PG&E plan] puts us in a position that redirects our focus in a more broad way; instead of just focusing on utility solvency, we ask what do we want the electricity system to look like in this state. What's the future of competition?"

PG&E's reorganization plan was to be the subject of a contentious dispute between state regulators and PG&E officials, as the plan would shift billions of dollars in assets out from under state regulations. Those familiar with utility regulations questioned whether PG&E had the right to transfer any facilities. In addition, utility commissioners had ruled repeatedly that public utility customers had a well-established right to a share of its assets. Robert Burns, senior researcher with the National Regulatory Institute in Ohio, remarked that PG&E's customers might be getting shorted if the transfer price was too low or if a big chunk of money from the sales was not applied to lower rates. The Institute, which advises utility commissioners nationwide, pointed to a seven-year-old study that found in most cases involving the sales of utility property, the profits from such sales go back to the customers in the form of lower electric rates. In other cases, the proceeds were split between the shareholders and the customers. In a minority of cases the shareholders received all of the proceeds. Burns said gains from operations are usually split between the shareholders and the customers. A basic question was raised, why should a utility customer "own" a piece of utility facilities, when a dairy customer can buy the same brand of milk for years and not own part of the dairy? Burns's counter-argument was "the ratepayer has no choice but to pay for the asset" because the utility is a monopoly. Regulators have long allowed utilities to collect from their customers all the costs of building needed facilities plus a moderate profit and when the asset was sold, the customers who paid for those facilities were entitled to recoup part of that investment. Burns went on to say, "It's not a legal ownership. But it's kind of a recognition that a majority of state commissions have done."

October

Attorney General Lockyer filed PG&E bankruptcy claims worth $1.1 billion on the behalf of a number of state agencies, including DWR, Department of Fish & Game, Toxic Substances Control, various regional water quality control boards, the Board of Equalization, and the Franchise Tax Board. Their claims ranged from unpaid taxes to environmental cleanup costs. Lockyer had previously admonished state agencies to avoid filing claims, believing such filings could jeopardize the State's sovereign immunity. However, the bankruptcy judge had advised that the State risked not being included in the bankruptcy settlement if it missed the October 3 date for submitting claims.

On October 2, the PUC and SCE reached an agreement in which both were confident they would keep SCE out of the bankruptcy court. The agreement would have allowed SCE to collect an estimated $2 billion in accumulated debt from the ratepayers. SCE would in turn be required to contribute more than $1 billion, most of which would come from suspended dividends. Rates would not increase for the ratepayers; however, the agreement would suspend any rate reductions that might be available in the following year. While it was highly questionable that the agreement would satisfy all of SCE's creditors, it gave SCE a window during which to both negotiate "mutually acceptable" agreements with creditors and avoid filing for bankruptcy. Harvey Rosenfield of FTCR denounced the deal saying, "This is the ultimate deregulation disaster—in approving this deal, the governor's un-elected appointees did to us what our elected lawmakers have refused to do for the last ten months—stick residential and small business ratepayers with the cost of keeping afloat a greedy utility company and its top executives." Rosenfield had threatened to fight at the ballot box any deal considered unfair to the ratepayers, but he recognized that since the agreement was reached through the federal courts, his efforts at the ballot box could be blocked. A relieved Davis applauded the PUC and SCE and rescinded the call for a third special legislative session. Earlier in the year, the PUC had granted SCE two rate increases, and SCE was now taking in more money than it needed to pay its operating costs. Assemblyman Keeley said, "They've been hoarding the money; they're going to use the money they hoarded and money they continue to collect." He did acknowledge the agreement was preferable to SCE's bankruptcy, as it avoided a reorganization plan similar to PG&E's that tried to remove hydroelectric and nuclear plants from under State regulation. It was observed that the agreement was remarkably similar to the rate stabilization plan the two utilities had proposed in November of 2000; the question arose, why couldn't the agreement have been made eleven months ago? PUC president Lynch opined that the current agreement gave SCE less and still kept power plants and power rates regulated by the State.

The PUC, by a vote of four to one, rejected a rate agreement proposal. Its position was that the proposal would have locked the ratepayers into needlessly high electrical rates for years to come. They recommended instead that the governor approve the legislation as proposed by Senator Burton that outlined a different structure for the power bonds. Davis called the PUC vote "an irresponsible act," which created uncertainty for the success of the State's bond sale. In addition, he vowed to veto Burton's bill, SB 18xx. Burton's response was "I would think the governor would look closely at the wisdom of vetoing the bill, given the PUC action. If he vetoes the bill, he's putting the State and himself into a pickle." The State was facing a $9 billion deficit if the general fund wasn't paid back by fiscal year 2002-03. After the PUC vote, State Treasurer Angelides opined that the PUC was "jeopardizing essential programs such as education, health care, and law enforcement." Several commissioners stressed that while they did not want to stand in the way of repaying the State; they just disagreed with the tactics. At issue was an estimated $1 billion in interest over the life of the bonds. The business community and consumer and environmental groups supported Burton's bill, which passed the senate 27 to 9 and the assembly 59 to 5, providing electrical rates that would repay the bondholders first. By contrast, a rate agreement rejected by the PUC would have paid the power sellers and then the bondholders. Critics and the governor feared Burton's bill would set the stage for the State to be sued by the power suppliers because of contract clauses assuring that the producers would be paid first.

On October 4, the CEC adopted "The Natural Gas Infrastructure Issues Report," which indicated there would be adequate supplies of natural gas for the coming winter. Conservation and efficiency had dampened electricity demand, with a corresponding reduced demand for natural gas, but sufficient storage of gas was the critical factor. Sufficient storage had been a critical problem for California, since 85 percent of the state's needed supply came from outside sources. With increased gas demands in the winter, interstate capacity had been inadequate without any backup supply normally provided by adequate storage. The report recommended that the CEC continue to access storage levels and monitor system improvements. The study acknowledged that California's natural gas systems had been designed many years ago, well before natural gas became the fuel of choice for electrical generation. The report analyzed available storage requirements and the capacity of interstate pipelines, which were deemed inadequate to accommodate the supply requirements. At the time, PG&E and Southern California Gas Company were in the process of improving their intrastate pipeline systems, and according to the report, their planned expansion was to be completed in 2002.

The PUC voted three to two on October 17 to remove the requirement that quick-building power producers cease their operations within three years or convert their systems to burn much cleaner fuel. This vote circumvented Davis's emergency rules passed by the legislature that would have allowed power plants to operate dirty "simple-cycle" turbines throughout their thirty-year life. The PUC also waived, temporarily, the rule that excluded developers of major polluting generators, i.e. those typically emitting one hundred tons of pollutants a year, from the agency's fast-track licensing process. The waivers applied to projects that had their license applications completed by December 19 and that would be operational by the following September. The CEC said there would only be a few that would be able to take advantage of the eased restrictions; however, environmentalists said about twenty projects could qualify. California Power Authority chairman David Freeman discounted their concerns, saying the simple-cycle plants would replace the older, dirtier plants. Although California's energy crunch had been relieved due to conservation, even with the addition of new power plants and long-term contracts, there still was still a need for more energy before the next summer in order to avoid price spikes and blackouts. The state power authority had lined up an additional thirty-two gas fired power plants capable of running up to 8,000 hours a year, all of which had signed "letters of intent" to build plants. The simple-cycle turbines were faster to build, taking only four months, compared to years for conventional combined-cycle plants. These were Davis's administration's generators of choice for reaching a goal of a 15 percent power surplus by the following summer. The CEC said the PUC order allowed the suspension of all restrictions in fast-track reviews that could prevent, hinder, or delay the permitting of power plants. Air pollution regulators, conservationists, and the legislature felt they had been ignored and wanted the PUC order rescinded.

On October 18, U.S. Secretary of Energy Spencer Abraham announced a plan to relieve the congestion between Southern and Northern California: "Path 15 has plagued Californians for over a decade and was even responsible for two blackouts last winter; we are taking a major step toward a solution that will relieve the pressure on this choke point in the transmission grid." The director for strategic planning at the PUC said, "This involvement of private, unregulated entities is troubling. The State of California expressed concern to the secretary sometime ago." The concern was that the owners of the transmission line would have the ability to restrict access to the line and use it to drive up prices. The PUC felt it should be allowed to review the ownership contracts. The PUC had previously asked PG&E to submit a plan for upgrading Path 15 and to commence an environmental review. The federal government would spend $300 million to construct the third line in Path 15. Construction was planned to commence in the spring of 2003, with completion in 2004. The owners would be a consortium of PG&E National, Kinder Morgan Power Company, Williams Energy Marketing and Trading Company, Trans-Elect, and various municipal utilities. Shares ownership was still to be negotiated; preliminary talks called for WAPA to control 10 percent and another 45 percent whose membership included SMUD and various other municipal utilities and electrical co-ops. Builders would own the rights to use the line and would be able to charge fees to other users, fees that would ultimately be paid by the ratepayers, but which would be regulated by FERC. It was unknown if that portion would be spun off to the new proposed PG&E company that had been in PG&E's bankruptcy reorganization plan.

On October 29, the U.S. Supreme Court ruled PG&E could proceed with a limited suit against the State, but it passed up an opportunity to resolve the limits and a number of legal questions that could arise once the State filed claims during the bankruptcy proceeding. California waived its right of sovereign immunity so it could claim $1 billion for energy purchases and other expenses.

In a hearing, U.S. district court judge Ronald Lew approved the rescue package negotiated by the PUC and SCE. The deal anticipated SCE would pay off its debt by the end of 2003; however, SCE planned to line up bank loans and pay off the debts by the end of March 2002. Edison International's chairman John Bryson planned to negotiate both with creditors and, in particular, energy producers, in an effort to reduce its debts. Mirant and Reliant Energy protested and asked the judge to delay approval of the agreement. Consumer advocate Rosenfield promised the decision would be appealed to either the U.S. court of appeals or through a voter initiative. Mike Florio of TURN asked the judge to reject the settlement, arguing the PUC had given SCE more than it needed.

On October 30, a federal appellate court put the State's bailout of SCE on hold and told the Los Angeles judge who had approved the deal to consider staying his decision until appeals were concluded. In granting the stay, the court opined that if the federal judge subsequently decided against the stay, the court of appeals might consider one. The recommendation did not guarantee a permanent stay or predict the PUC's bailout would be voided.

November

The electrical energy crisis brought the champions of public power out in full force. UCEI's Borenstein observed, "The case has not been made that munies are the answer. I don't think most cities are equipped to do this. It is doubtful they will find a great deal that somehow the private sector has missed." Jan Schor, general manager of SMUD, disagreed, saying, "You can build new resources at competitive prices. We're doing it." (A listing of the state's utilities: public, private, and cooperatives, is contained in Appendix A.)

Mirant and Reliant complained to FERC that California's ISO had been providing bidding information to the DWR that was not made available to private generators. FERC in turn ordered the ISO to stop giving preference to power purchased from the DWR. The concern was that confidential bidding information, such as availability and demand, gave preference to the DWR, even though it may have been more expensive. The DWR argued it was helping the ISO find cheaper power than would have been provided through the bidding process, in the interest on keeping prices down. The ISO pressed FERC to dismiss the complaint, arguing DWR's status entitled it to non-public information, but FERC rejected that argument. One consequence of FERC's order was that DWR's power sales came under the agency's tariff regulations. The House Energy and Commerce Committee sent Governor Davis a letter demanding documents and information about the DWR and ISO dealings. FERC had been concerned that the DWR, under the governor's control, strongly influenced the ISO, claiming that the relationship hampered the theory of free markets. A related matter existed over a DWR billing dispute of $1 billion of energy purchased through the ISO. The DWR wanted the bills to name suppliers, but the FERC said the ISO wasn't allowed to put such market-sensitive information on its invoice to the DWR. The DWR spokesman argued, "We need the basic information that any responsible entity needs to know before dispensing funds . . . who are we paying, how much and for what."

Enron had been the product of a merger of two natural gas operators when the federal government began opening up the energy markets. Enron seized the opportunity to become an aggressive leader in trading gas and electricity. In the 1990s, Enron was actively lobbying the federal and state governments to open up the electrical markets to deregulation, as it favored a wide-open market for the buying and selling of energy. Enron was influential in having California deregulate, but the State rejected Enron's "wide-open market" for a centralized venue market. When California started deregulation in 1998, customers were allowed to select their electrical energy suppliers. Enron had aligned itself with several prestigious accounts, such as the University of California system, but quickly moved out of the retail market and concentrated on the wholesale market. The state's utilities, as required by AB 1890, sold off much of their generating capacity, thus relying on the wholesale market, including Enron, for its electricity. In the year 2000, Enron's sales in California totaled hundreds of millions of dollars; its total sales were $100.8 billion, earning a profit of $979 million. In the first six months of 2001, its sales were only $44.3 million. Enron's end began in October when it posted a quarterly loss of $618 million. Additionally, Enron reported a $590 million error in reported profits in 1997, saying it was an accounting problem. Dynergy had proposed a takeover of Enron but bowed out after these revelations.

An analysis demanded by the legislature and produced by the DWR revealed what California's energy costs were projected to be in the ensuing years. The study revealed that Davis's long-term "take-or-pay" contracts would exceed $4 billion on a total portfolio of $39 billion. The terms of the contracts left the DWR in an untenable position. On the one hand, they could sell only the energy demanded of it but if a shortfall occurred, the DWR would have to eat the difference between the contract amount and the shortfall.

Attorney General Lockyer filed a complaint against PG&E's reorganization plan in the bankruptcy court, claiming the utility was using the bankruptcy process to make profits for its parent company by selling its hydroelectric facilities to them. The State further contended PG&E was selling these assets to its parent company for a price that was half of its market value. The State also maintained PG&E was using this ploy to remove itself from under the state's environmental regulations and its long-standing agreements. The State requested that PG&E identify all laws and regulations that would be preempted by their plan. PG&E countered that it was up to the State to identify the laws and regulations it believed would be preempted.

December

The PUC voted five to zero to undo its November vote that would have allowed peaking power plants to run longer without upgrading to a cleaner configuration and that would allow the major polluters to use the faster licensing process if the plants applied by December 19. Air quality officials and environmentalists were grateful to the PUC for paying attention to their concerns. The PUC pledged to move quickly on the three plants that would have been affected by the streamlined process. Calpine had a plant that would have been affected, and the vote caused the project's expected completion date to slip to 2003. The 135 Mw power unit was to be put alongside the existing power units next to the garlic processing plant in Gilroy, California.

Nevada entered the energy fray after having experienced a crisis similar to California's. Nevada Power Company and Sierra Pacific Power Company filed formal complaints with the FERC, arguing that their long-term contracts were products of a time when spot market prices were so far out of sight the FERC had to impose price caps after finding the contract prices were "the product of markets to be dysfunctional and not competitive," and concluding that "the dysfunctional spot markets directly impacted the prices in the forward markets [long-term contract prices]." The complaints were against ten energy suppliers, including some that had contracted with California. Energy suppliers didn't believe the Nevada utilities would be successful. However, some experts believed California should consider the same leverage tactics in trying to renegotiate their contracts. Governor Davis and others had been talking about renegotiating the $43 billion worth of contracts since July.

State audits had been performed and rushed to completion on the $43 billion worth of long-term contracts that had been negotiated by the DWR. The audit stated, "Although the department was in a weak bargaining position, its rush to ease the electricity crisis by locking in power supply through long-term contracts weakened its position even further." Pete Garris, DWR's acting deputy director, admitted the terms and conditions could have been better but felt the audit "missed the point." He went on to say, "We were given the responsibility to get reliable, reasonably priced energy, and . . . we've influenced the market enough where I think our energy is reliable and as reasonably priced as it could be." The audit singled out three energy contracts that were particularly suspect—those with Calpine, Williams, and West Coast Power. These contracts had been negotiated in secret but were made public at the direction of a court. The contracts shifted some energy costs to the buyer. And in two of the agreements, the DWR agreed to relieve Williams and West Coast of hundreds of million of dollars in polluting costs and fines for air pollution credits that allowed power plants to exceed permitted emissions. Bill Marcus of JBS Energy observed, "It creates a perverse [ability] to simply pay to pollute rather than clean up." A number of the first agreements bound the State to pay generators' costs via an array of governmental actions that included higher taxes on natural gas, increases in water rates, and various environmental costs. Auditors calculated that the State could potentially be liable for up to $688 million in air pollution costs under the Williams contracts. In grading the contracts, the auditors assigned the lowest marks of reliability to Mirant and Calpine. Marks for the worst in assurances on building new generation went to Calpine and Sempra. In addition, the DWR hadn't included clauses providing for cancellation and/or no delivery penalties. The audit advised the DWR to employ energy experts and lawyers in order to assure that the contracts would be enforceable to the benefit of the State.

PG&E identified a list of state laws, as had been requested by the bankruptcy judge, which would be unenforceable by the State if their reorganization plan had been approved. The list included more than thirty statues. In revealing the list, PG&E gave assurances that the plan would not involve increased rates or open its extensive land holdings to unregulated development. PG&E also assured that 70 percent of its assets would remain under control of the PUC.

V. Midnight—Year 2002

January

On January 2, PG&E asked the bankruptcy judge to issue an order against the PUC's Office of Ratepayers Advocates to stay the PUC's requests for information on PG&E's financial relationship with its parent PG&E Corporation. On April 3, 2001, the PUC voted to proceed with an investigation of PG&E's transfer of $4.6 billion to its parent over the past five years. PG&E argued it could not be forced to turn over the documents requested except in a bankruptcy proceeding. They further argued that the release of such information would interfere with the bankruptcy process and that the PUC might use the information to lobby against its reorganization plan. The PUC believed it was acting under a statutory mandate "to find out what the utilities are doing and what impact of any of their proposals might be on the consumers."

The PUC released its own plan for resolving PG&E's bankruptcy. The plan would have forced PG&E to pay its bills before it paid its attorneys. The PUC charged that PG&E's plan was an effort by PG&E not to pay its debts and to remove itself from the PUC's oversight. The PUC observed that the outrageous wholesale prices that had driven PG&E into its enormous debt no longer existed, as wholesale prices had plummeted. The PUC plan would also force PG&E to pay its thirteen thousand creditors from its excess revenues (which were almost $5 billion) and would include in making any such payments refunds it might receive from its parent company and its creditors. PG&E had an exclusive right to delay proposing a reorganization plan until a February 4 deadline and asked the bankruptcy judge if the deadline could be extended to June 30. The PUC, in opposition to the extension request, promoted its alternative proposal. A hearing for the extension motion was set for January 16. If the judge ruled against PG&E, the PUC was immediately ready to formally proceed with its plan. The PUC proposal, while supported by both California's attorney general and the bankruptcy trustee, stopped short of endorsing the PUC's formula. The attorney general had been challenging PG&E's plan, which would have relieved PG&E of conforming to state regulations. The representative of the U.S. Department of Justice, which oversaw the case, stated she would file an objection to the request for the extension. The FTCR criticized the PUC proposal as "an unfair bailout" for the utility that wouldn't give ratepayers immediate reductions.

Governor Davis proposed reducing the biomass grant program from $11.5 million to $6 million as part of his spending cuts. The San Joaquin Valley's biomass operators believed that without the $11.5 million grant, they couldn't afford to burn much of the millions of tons of pruning, vineyard, and orchard remnants, all of which require disposal during each farming season. Without the grant, the excess waste would have to be burned in open fields, impacting both the environment and air quality. The Valley had been ranked among the six highest places in the U.S. for dust and soot air contamination and had been included in the eleven worst places for ozone and smog pollution. Peter Weiner, lobbyist for the California Biomass Energy Alliance (CBEA), said, "This is a significant hit at [a] small but vital part of the fight for clean air quality in the Valley, as it is economy." He went on to say, "There could also be 300 jobs at stake when you look at the multiplier effect in the Valley, from the people who grind up the wood to those that haul it to the plants." The waste consumed by these generators has been one of the most expensive fuels used because of the high cost of grinding and transportation. These operators can contribute 700 Mw to California's grid; however, without the grant program operators could not make enough from sales to cover their production costs. It was both of these concerns, along with air quality, that produced the grant program in 2000. The three-year $30 million program had revived several struggling and closed biomass power plants; however, all of the $10 million that was available in the first year had been used. Kent Duysen, an 8.5 Mw plant operator, said, "Everyone expected to be fully funded—everyone's gone out and procured contracts, and now it looks like we'll be left holding the bag." Davis's spokesperson said it was because the operators hadn't submitted their grant applications for the current year, before Davis had announced the cuts. State officials believed the remaining $5.5 million of the program could be reverted to the general fund without hurting the industry.

California's yearlong drive to build more power plants appeared to be slowing down. Completion of two of three large power plants was in flux, and there was no certainty as to just how many additional power plants would be built. Steve Larsen, executive director of the CEC, said, "At what point do companies decide prices are low enough that they don't want to build any more? I don't know how to forecast that." Three major power plants and about a dozen smaller ones had been built and/or started up in the past year and there were many more large power plants in progress, eleven of which had been approved and fifteen that were in the licensing process. A FPL spokesperson commented that not all of those plants would be built, as "they're not economically viable." The CEC predicted that by 2004, California would have a power surplus. CEC's Larsen expected all of the twenty-one power plants that had been licensed to be built would be built, but industry executives believed about half of those plants was more realistic. Once the 750 Mw plant in the city of Roseville was permitted, Enron was expected to sell its license. Calpine faced pressure from its shareholders and lenders who were concerned about Calpine's financial condition. Doubts were raised regarding Calpine's ability to double the amount of kwh it produced in California. FPL put on hold its plans to build a 560 Mw plant in Rio Linda and asked the CEC to suspend its application. However, SMUD was preparing to build a 500 to 1,000 Mw plant at its defunct Rancho Seco Nuclear Plant site.

On January 16, Calpine put a screeching halt to building its proposed power plants, putting billions of dollars in power plants on hold, including 4,800 Mw of plant capacity in California. However, it remained committed to completing every power plant that it currently had under construction and to maintaining its long-term plans for constructing new power plants, although that construction would proceed only if Calpine could obtain financing and if there was enough demand for electrical energy. State officials were concerned that a building slowdown would result in a repeat of the crisis as the economy recovered, resulting, once again, in an increased demand for electrical energy. Power Authority chairman David Freeman commented, "The power plants are dying like flies out there—how dumb can we get? It's pretty clear that shortages mean higher prices.

These folks have every incentive in the world to keep us on short rations. As soon as prices come down, they lose interest in building power plants." Peter Navarro, professor at UC Irvine, observed, "The problem with deregulation is you get boom or bust." Other experts observed that California, for the foreseeable future, as a result of entering into long-term contracts, had little chance for a major shortage anytime soon. The DWR had predicted the State would end up paying for large amounts of power and selling that power at discounted prices with a loss of billions of dollars. Borenstein said, "Companies like Calpine are merely putting on hold projects that would have made sense only if prices had stayed sky-high. I wouldn't have been surprised if these guys had plowed ahead—then there would have been a real glut." In a hearing, the PUC was concerned over other power projects being put on hold—FPL's 560 Mw plant in Rio Linda and Mirant's 540 Mw plant in San Francisco. In addition, Enron's bankruptcy left the Roseville project in limbo. Overall, there were concerns that if too many plants were canceled, California could be headed for another energy crisis.

In California there was a serious debate regarding the fifty contracts totaling $43 billion for electrical energy that the DWR had negotiated with energy suppliers. The objective was the deletion of the most egregious terms from those contracts. The power producers maintained they had fair and valid contracts and would discuss only changes that were mutually beneficial. Michael Shames, head of UCAN, said, "The contracts that are in place right now will serve to keep energy prices paid by almost all Californians excessively high for the next seven to ten years." He went on to say that manufacturers, growers, retailers, and countless others would have to compete with rivals in neighboring states that have far lower energy costs. However, Calpine, Dynergy, PacifiCorp, Sempra Energy, and Williams all confirmed that they were in negotiations with the State. In addition, Constellation Energy and Allegheny met with the State, and Coral Power was scheduled for talks. Consumer and environmental groups criticized the contracts involved in these negotiations, as they happened to be largest power contracts. The State sought to revise portions of the contracts that dealt with lower prices, reliability of power, and assurance that power deliveries would match the State's requirements. Paula Hall-Collins, spokesperson for Williams said, "Our position has been clear from the very beginning. We do have a contract, and we feel it was fairly negotiated. But as with any of our long-term deals, we're interested in developing a relationship [when] environments change. When something is not beneficial to either party, it's time to sit down and talk." On California's side were ten representatives of Davis's office, representing the DWR, the PUC, and the attorney general's office. Power marketers, at the time, were having their troubles, as stock prices were falling and lenders had become reluctant to extend lines of credit. It was questionable whether the power marketers would consider the possibility of having to pull out of their contracts. At the same time, there were concerns that many of the contracts in question were tainted by conflicts of interest or other improprieties. State Auditor Elaine Howle, who issued an analysis of the contracts, told the Joint Legislative Audit Committee that many of the contracts lacked "meaningful" avenues for either negotiation or termination. She recommended a new legal team be assigned to review each contract. She opined such a review should look for leverage to encourage sellers to accept better terms for the State. She said, "Our consultants believe a couple of the Calpine contracts were two of the worst contracts." The audit gave Williams, Dynergy, and Sempra low grades on reliability. UCAN and other groups believed the worst contracts were those that required the State to pay a fixed amount for power even if the power wasn't used (take-or-pay contracts). There were six contracts with five parties that were the most troubling, and those contracts accounted for 60 percent of the energy cost and two-thirds of the power that the State would buy.

Many questions arose regarding the future electrical energy situation. Lawmakers, regulators, and others decided that now was not the time to take up the pending "nagging issues." The DWR's emergency authority was scheduled to terminate on December 31, 2002, raising the question of who would buy power after that. There were also open issues of renegotiating long-term power contracts as well as the sale of the long-delayed energy bonds. Other major questions existed, such as, would the three large utilities be ready to assume buying wholesale power again, and how would the existing long-term contracts be distributed among the utilities? Some of the contracts contained terms more favorable than others, so who gets what power and when? PG&E felt it would not be ready to purchase power until after the first quarter of the next year and then only if the bankruptcy court limited the reach of the PUC. The ISO also was concerned with supply issues and started revamping its trading rules in an effort to create and make its short-term purchases on a day in advance basis. They also continued to lobby the legislature, attempting to help lawmakers get a better understanding of the ISO and how it functioned. Some lawmakers wanted to better understand the ISO in order to take a hard look at its management. Senator Bowen wanted to formulate an energy conservation plan and determine a billing plan that would accurately charge large customers for power they used during the high-price hours. The California Business Roundtable and the Chamber of Commerce recommended consolidating the PUC and the CEC under a new state energy department.

After sitting on their hands for four months, the PUC finally reversed itself on the rates the three utilities should pay the DWR for energy; a final vote was expected to come in February. The new proposal called for PG&E and SCE to pay slightly more than 10¢ per kwh and SDG&E to pay slightly less than 10¢ per kwh. The payment formula would apply from January 17, 2002, through the end of that year.

PG&E filed a $4.1 billion claim with California's Victim Compensation and Government Claims Board on January 17. It claimed AB 1890 allowed the utilities to sell power from their remaining nuclear and hydroelectric plants at market prices. In PG&E's view, the State had reneged on that promise when it passed AB 6x last year.

On January 17, 2001, the state was plagued with its first rolling blackouts, and Governor Davis declared a statewide emergency.

In two draft decisions released on January 18, the PUC said PG&E should change the way it charges for nuclear power and peg its other charges to cost. It spelled out how much PG&E should bill its customers for power produced or purchased under contracts. The draft, if adopted, would allow PG&E to collect $2.9 billion for generation costs in 2002; PG&E had sought $6.4 billion. The issue was scheduled for a full PUC vote on February 7. Nettie Hoge of TURN said the PUC proposal provided figures low enough to prompt rate cuts. PUC president Lynch opined that a PG&E rate cut was likely in 2003, but there was no speculation on how much that cut would be. The first draft decision clarified how much the utilities should pay the DWR for power. A second decision calculated how much PG&E could pay itself for power from its own nuclear and hydroelectric plants and how it should pass through QF's contract costs. The PUC proposed PG&E abandon an incentive plan for its nuclear plant and rejected the utilities' arguments that it should replace the incentives with a profit-sharing plan. The PUC instead recommended that

PG&E be allowed to collect its costs plus a 9 percent profit, totaling about $440 million a year, as opposed to PG&E's proposed $550 million a year. In a side comment, apparently PG&E sought about $1.3 billion from its profit-sharing plan.

PG&E and PG&E Corporation had presented a new Plan of Reorganization and Disclosure Statement to the bankruptcy judge on December 19. The PUC objected to the plan and filed its opposition with the court. Both of PG&E's September and December proposals called for PG&E and PG&E Corporation to be two separate companies. PG&E Corporation would divide into three subsidiaries: Electric Generation, Etrans, and Gtrans. Electric Generation would be responsible for generation, including hydroelectric and nuclear generation; Etrans would be responsible for all aspects of transmission and substations; and Gtrans would be responsible for natural gas transmission pipelines and natural gas storage. All three would be regulated by the FERC while PG&E would remain an investor-owned utility providing retail electricity and natural gas to all its Northern and Central California customers. There was no anticipation of any rate increases in the plan, and PG&E would continue to be regulated by the PUC. Any changes between the September and December submissions had been the result of objections made mainly by creditors. If the bankruptcy judge accepted the plans for reorganization and the proposal contained in the disclosure statement, the bankruptcy proceedings would end by December 31, 2002. The PUC was concerned about hidden rate increases in the plan, stating:

> The CPUC suspects PG&E's plan is premised upon a disguised rate increase. PG&E's plan and disclosure statement may be premised upon the assumption that a 3¢ increase ordered by the CPUC in March 2001 belonged to PG&E, but that remained unclear, as a portion of that 3¢ increase was to be allocated to the California Department of Water Resources. If PG&E kept all of the increase (or any portion that would otherwise belong to DWR) then its electricity rates must necessarily increase. In addition, PG&E proposed, in its bankruptcy proceedings, to sell its retained generation at rates higher than the cost of service.

In its disclosure statement, PG&E asked the bankruptcy court to allow preemption, on a one-time basis, of the thirty-seven PUC regulations.

They argued that the court had the authority to override any laws deemed non-applicable in order to allow the debtor to take actions necessary for it to emerge from bankruptcy. PG&E in its statement went on to argue that in order to complete the transfer of certain assets, and unless the thirty-seven PUC regulations were preempted, they would be restricted both in the transfer of property and the issuing of new securities. These measures were needed to implement the reorganization plan and restrict the execution of key contracts. The PUC continued to maintain that PG&E's objective was to free itself entirely from PUC oversight. FERC's approval would be needed on any transfer of assets and/or contracts between the new companies. PG&E's request included immediate settlement of all claims under $100,000. On January 22 and 25, PG&E had been scheduled to file its responses to the PUC's briefs on preemption and sovereign immunity issues.

PG&E, DWR, and PUC exchanged lists of potential mediators in an attempt to find people who could bring some resolution to the bankruptcy turmoil. It was anticipated that in the following week a mediator candidate, acceptable to the three parties, would be found and be appointed by the bankruptcy judge. The judge had voiced his opinion that he favored the mediator to whom PG&E had originally objected, and PG&E finally relented and agreed to cooperate. PG&E still had reservations about the delays to its reorganization plan. The judge heard arguments on January 25 on whether the bankruptcy proceedings could be used to sidestep thirty-seven state regulations on retail rates and environmental issues. The judge didn't know how and when he would rule on that issue. The PG&E plan relied on the assumption that if the controls were relaxed, it would make it easier for PG&E to borrow money, while the PUC was concerned that under PG&E's plan; public health and safety would be jeopardized.

The bankruptcy judge came to the conclusion that he could not let PG&E use the bankruptcy process to remove itself from state regulations. He asked the PUC to expand on the outline of its plan. The judge, extending the February 4 deadline, gave PG&E until February 13 to present a revised plan, but PG&E had asked for a deadline extension until June 30. The judge characterized PG&E's plan as "enormously risky" and was receptive to the PUC's objections to PG&E's requested extension. He stated, however, that he needed the PUC to show him a really viable alternative. He wanted to know specifically where and when the money would come from and which creditors would be paid. If the judge agreed that the PUC plan was workable, then he would circulate the plan to the creditors. The judge also said he could consider the possibility of mediation, and he requested that PG&E and the PUC recommend neutral mediators who might come together with a solution. The PUC had observed that PG&E had $5 billion on hand which would increase with future revenues and that PG&E could pay its creditors from those funds.

The California PX, joined by the State, filed an emergency petition with FERC seeking help to keep power marketers from their planned takeover of PX's books and records during the pending bankruptcy proceeding. The PX charged that if the marketers gained control, there was the possibility they would immediately refund to themselves approximately $1 billion in collateral which the sellers had posted, in order to assure power deliveries. The PX management argued that anything paid to the marketers would drain cash that could be used in any federal remedy FERC might adopt. In addition, the petition asked that PX's books and records not be given to parties in the dispute currently under investigation by FERC. Meanwhile, a "participant creditors committee" filed a bankruptcy reorganization plan in December 2001 that sought to take over the management of the PX. The petition claimed the PX owed members of the committee about $2.5 billion. Members of the committee included Automated Power Exchange, Avista Energy, LADWP, Mirant, Americas Energy Marketing, Powerex Corporation, Reliant Energy Services, and Sempra Energy Trading Corporation.

FERC was looking at organizing multistage transmission grids to be regulated in Washington DC. It essentially wanted to create four major regional transmission organizations (RTOs) to run the transmission configuration in the continental United States. FERC believed this reorganization would lead to more competition, boost system efficiency, open competition, and that it would determine how the electrical industry would emerge nationwide. Those that favored state-controlled grids worried that the larger buyers would receive special favor. State Senator Joe Dunn (D-Garden Grove), chairman of a committee investigating price manipulation, said, "What's in play here is an effort to totally deregulate their industry, in all phases of it, so they can basically do as they please." He went on to say that California must oppose a regional grid until the FERC can prove they have truly conquered every effort at manipulation of the market which forces prices up. Nine states, including New York, had argued that FERC was going too far and would be intruding into retail transmission that had been under the purview of state regulators. FERC chairman Pat Wood declared that New York's position was virtually irrelevant and that if the court tried to limit FERC, he would take the issue to Congress. California, in a friend of the court brief, added its support to the other states, stating that FERC had severely limited California's alternatives in its crisis and pointing out that California had an unresolved dispute with FERC, including who had the authority to appoint the ISO. Advocates of open markets distrusted the ISO board, as it appeared to be a tool of Governor Davis; it was formed after the governor had forced out the original ISO board and converted it to his five-member governor-appointed board. States in the Northwest and Southwest had begun to create their own transmission organizations; however, they were wary of California; no one wanted California included in its RTO, as they felt there would be problems in trying to deal with California. FERC recognized the reluctance and in November issued an order saying the West might be divided into three small regional grids under a larger RTO. But at the same time it didn't abandon its goal of a larger organization.

FERC ordered an audit of the ISO operations. After the audit, FERC recommended that the existing five-member board be replaced by a seven-member board with three members chosen by the current board and three chosen by a stakeholders group. The seventh member would be the current ISO Chief Executive Terry Winter. In 2000, the FERC had ordered that a new management-selected ISO board be chosen. But the state legislature had authorized a governor-appointed board. The auditors in their review had interviewed officials in state agencies, utilities, generators, marketers, FERC, California's ISO, and ISOs from other states. The conclusion was, "With the exception of other state agencies, none of the stakeholders' other parties we interviewed believe that the ISO could ever be truly independent under this arrangement."

The governor, the state treasurer, and the PUC finally crafted a plan that would allow the State to move forward in issuing up to $12.5 billion in bonds. The PUC had objected to the plan, as it would have guaranteed that power marketers would be paid before the State. The new plan did not sit well with consumer groups, who felt the plan abandoned efforts for stricter controls over power spending and offered little protection for ratepayers. PUC president Lynch backed the plan; in her opinion, circumstances had changed and the plan was the best they could do. A draft of the plan was released on January 31, spelling out how the ratepayers' money would be distributed between the DWR and the bond investors. The draft outline gave the PUC more funding than the previous plans allowed; it provided more data regarding electrical costs and reserves; and it called for the best efforts of the DWR to renegotiate the long-term contracts.

February

With the release of the draft, a number of the questions were clarified, but several issues remained to be addressed before the bonds could be issued. Still to be resolved were any court challenges and a legal opinion stating the plan was valid. The FTCR group was concerned the plan would "remove the public oversight of energy costs." PUC Commissioner Richard Bilas said, "We're getting precariously close to the point where if we don't issue bonds pretty soon, the state's going to be in very serious trouble."

Republican Congressman Doug Ose convened a congressional hearing in Sacramento to probe the future of the state's energy market, including the complaints outlined in a FERC report. He stated the ISO had become a tool of Governor Davis. Terry Winter, ISO chief executive, testified that California's electric market would need permanent price caps in order to guarantee that the state would never again face widespread runaway costs. He went on to say he would soon ask FERC to approve two new sets of cost controls and an extension of the current FERC price cap, which was scheduled to expire in September. FERC chairman Pat Wood predicted the FERC would agree to scaled-back cost controls rather than allow prices to float freely. Opposition to the extension of price caps was expected from the power marketers, who argued that price controls thwarted the price run-ups that encouraged the industry to build power plants. Jan Smutney-Jones, IPPA, said, "Price caps should be imposed only under truly extraordinary circumstances, if they're applied at all." Power producers seemed to recognize that some sort of price pressure relief would be appropriate in "very rare circumstances." The ISO, in turn, was developing a plan to replace FERC controls with a "damage control" cap, somewhere under $1,000 per mwh, with a fallback safety trigger to further rein in prices if they moved too far from the twelve-month norms. Terry Winter argued controls wouldn't work unless the utilities were financially healthy enough to buy their own power.

FERC ordered on February 13 a nationwide investigation into wholesale electrical and natural gas markets with the focus on whether manipulation by Enron and other energy traders caused the outrageous prices in the past year. Pat Wood, FERC chairman, assured there would be "a full bore investigation" into both energy transactions and financial trades similar to those that had been dominated by Enron's online trading division. While Enron was the dominant culprit, wholesale power and gas trades of other energy companies were to be included. The investigation was expected to take six months and after a review of the staff's findings, FERC would decide if another round of investigations would be required. The objective was to determine if changes in the long-term contracts were in order, including whether Enron's activities had influenced the contracts. Senator Maria Cantwell (D-Washington) was disappointed that FERC had not moved more quickly to investigate whether some of the high-priced contracts had been improperly negotiated and whether they should be revised.

The PUC, on February 13, released its reorganization plan that was designed to get PG&E out of the bankruptcy court. The PUC plan included the following provisions: (1) PG&E would continue charging its customers more than it pays for power until the following January; (2) PG&E would not diminish the PUC's authority in rate setting and environmental controls; (3) PG&E would suspend dividend payments to its shareholders, allocating the payments toward $8 billion of creditors' debt; (4) PG&E would use the $5 billion in cash for payment toward $8 billion of creditors' debt; (5) the company would not depend on payments from the PG&E Corporation or refunds from power generators; (6) PG&E would depend on the company's ability to refinance $5.8 billion in long-term debt. The plan projected long-term savings of $13.3 billion to PG&E's consumers.

The three participants—PG&E, PUC, and creditors committee—agreed to a mediator and were awaiting the judge's order to commence talks. As expected, PG&E had a hostile reaction to the PUC's plan. PG&E said, "It will not work because it overstates the available cash, understates the debt, and undermines the utility's ability to invest in electric system reliability. Most striking is the fact the PUC continues to misunderstand the meaning of creditworthiness. The PUC's term sheet will not restore PG&E to creditworthy status and thus condemns the State to remain in the power business." The law firm representing PG&E's Unsecured Creditors Committee (UCC) released this statement:

> The Official Unsecured Creditors Committee ("OCC") in the PG&E Chapter 11 case just received the Plan Term Sheet ("Proposal") from the California Public Utilities Commission ("CPUC") and the State of California but has not yet had an opportunity to review or discuss the proposal. The UCC will carefully review and analyze all aspects of the proposal. The UCC expects to discuss the proposal with the CPUC, the State and PG&E. The UCC also looks forward to playing a constructive role in the upcoming Chapter 11 reorganization plan meditation process.

The parties had until February 21 to respond to the PUC proposal.

On February 20, PG&E and its parent company told the bankruptcy judge that the PUC plan was "patently defective" in that it contain a $4.5 billion error that could be covered only by a large increase in retail rates. They said the proposal overlooked basic obligations such as taxes and interest payments, and it failed to budget money for improvements in order to prevent blackouts and restore PG&E's creditworthiness. Restoring PG&E's creditworthiness, in their view, was critical, as it would enable them to start buying power again. The bankruptcy judge was skeptical of the PUC's basic idea of paying off the utility's $13 billion debt by using excess money that had been collected from the ratepayers and the suspending of dividend payments. However, he did agree to review the PUC's numbers. The court was concerned that PG&E's concept of transferring billions of dollars in assets out from under PUC control had serious constitutional problems. The PUC president said the major dissimilarity between the two plans was a philosophical difference on how the long-term debt should be handled. PG&E proposed paying off the debt much like paying off a mortgage early. The PUC, however, wanted PG&E to pay off the debt month by month as the utility collected revenues from its customers.

February 25, California officials filed a petition with FERC to be released from the $43 billion of long-term contracts that had been negotiated; they argued the prices in those contracts were unreasonable and had been executed under duress. Davis observed that at one time the FERC had placed price controls on California's power, and he believed they could help California again. Calpine's expectation was that FERC would uphold its contracts with the DWR. The state's Electricity Oversight Board joined the PUC in filing similar complaints of overcharges. The PUC's complaint listed thirty-three contracts with twenty-two marketers amounting to $43 billion arguing they were $17 billion to $20 billion above current prices. Not all were confident about the success of such complaints. Frank Wolak of Stanford said, "FERC certainly has tremendous discretion to reset or abrogate contracts . . . But they certainly seem to have no appetite to do so." UCEI's Borenstein said, "Their track record is to sit on these things."

The bankruptcy judge ruled that PG&E was required to explain to the court why they should be allowed to sidestep thirty-seven of California's laws and regulations. The decision rejected PG&E's argument that bankruptcy law preempts the PUC laws and regulations. Bankruptcy experts and a federal official opined that the decision would likely drive the issue to the mediation table. The creditors' committee legal counsel encouraged the PUC to present their explanations as scheduled by the court. In addition, the creditors wanted active participation in any mediation. The PUC, the state attorney general, and environmental and consumer groups argued that the splitting of PG&E would rob consumers of assets they helped pay for as well as remove needed controls.

California's newly created Power Authority found itself looking for a mission. Its role had been a contentious issue in the governor's race—the three Republican candidates questioned its existence, while Davis staunchly defended it. The authority, consisting of about a dozen members, had come up with several versions of how to spend $5 million in bonds, if it could find a way to issue them. The effort to build peaking power plants had terminated. The authority had trouble trying to find a buyer for its power. The authority could have helped build a single peaking plant at the San Francisco International Airport, if the PUC and the DWR had agreed that the utility customers would eventually pay for the plant. Another idea was the concept of peak shaving—recruiting businesses that would agree to shut down some operations during periods of peak demand. However, here the authority was again faced with the PUC and the DWR coming to an agreement that the consumer would pay for it. These agreements for the selling of bonds required demonstrating to the investors that there would be a revenue stream to repay the bonds.

Additionally, the PUC would have to be persuaded that the deals were good for the consumers.

March

March 1 was a notable day for SCE. Using a new $1.6 billion loan and cash on hand, SCE paid off $4.5 billion of its debts to lenders, power generators, the DWR, and various other creditors. These payments met the schedule SCE had set last fall in a pact with the PUC. The pact settled SCE's federal lawsuit against the State by giving SCE permission to gradually collect billions of dollars from its customers, thereby allowing SCE to settle old debts. Consumer groups abhorred the settlement, saying it reneged on deals made at the beginning of deregulation, and consumer advocates challenged the settlement.

FERC received a staff report on March 13 indicating that little had changed in California since the blackouts of the preceding year. However, there were two favorable contributing factors—California's economy and weather. Due to the recession, energy demand and prices were down, and a heavy snowfall in the West would assure adequate hydroelectric generation in the summer. The report indicated fewer power plants were being built than had been anticipated, and the state budget had cut money from energy conservation programs. FERC commissioner William Massey, the one commissioner who firmly supported California during the crisis, said, "I'm concerned about what happens after September. What happens when price mitigation comes off? Reasonable prices require good infrastructure and good market design, and we don't have that yet in California. I want to see much more progress in all these areas." Davis's press aid said while power generators were backing off from future construction, the State had approved a power authority to build its own plants. The staff report, relying heavily on information from the PUC and the CEC, stated that new power plant construction was lagging and 15,000 Mw of new power plants had been tabled or canceled. This represented a 47 percent reduction over what had been anticipated to come on line by 2006. The State's elimination of incentive programs to build new power plants along with reduced conservation activity contributed to the commissioners' anxiety. California got 40 percent of its summer electrical energy from hydroelectric power, and work was in progress to upgrade Path 15. However, there were indications that California would again be unable to meet its power demands from in-state production. At the same time, planning for new interstate power transmission facilities had been lagging. FERC commissioner Nora Mead Brownwell said, "We're relying on two very unpredictable aspects of the market—hydropower and cooler weather. We're assuming the recession in California lasts perhaps longer than is now estimated by economic forecasters, and we're hoping the transmission project gets through." Meanwhile, Davis was under attack by Republican leaders of the House Energy and Commerce Committee and in particular by chairman Bill Tauzin (R-LA), who stated, "The inadequacy of your response leads us to believe that you may be stonewalling the committee's review."

The PUC, on March 21, voted three to two, over the objections of PUC president Lynch, to allow large energy users to retain their direct-access contracts but require them to pay an "exit-fee" for the privilege. The category of "direct-access customer" was the result of deregulation, which allowed unregulated power producers to buy and/or build power plants and allowed consumers to shop around for their power. The decision caused the small ratepayers to protest strongly, as they believed they were being left to pay for costlier power purchased by the DWR; it also angered many Democratic legislators. TURN's Mike Florio said, "Every penny the big businesses save will be on the backs of the rest of us." Senator Bowen said, "I'm skeptical the PUC will be able to protect California's ratepayers from having another $4 billion to $5 billion in power costs loaded onto their backs." Large electrical energy users like the California university system, high-tech companies, cement, steel, food processors, and retailers welcomed the opportunity, while many residential customers generally opted to remain with their original utility companies. The question before the PUC was, how much should the exit fees be? One estimate was around $340 million a year, about 2.25¢ per kwh for all electricity used in the direct-access contracts.

In the year 2000, PG&E had posted a loss of $3.36 billion but turned around in the year 2001 to post a profit of $1.1 billion. This turnaround was largely the result of falling wholesale energy prices that began in the preceding summer when prices fell from more than $200 per mwh to within the $50 per mwh range. This fueled protests by consumer advocates, who felt that these profits were further evidence that the PUC should cut retail rates. The attorney for TURN, Mike Florio, said the PUC gave PG&E a big rate hike the preceding spring with the understanding that money would be refunded to consumers if wholesale prices fell: "Instead of that money being refunded to consumers, we now have PG&E reporting a profit." TURN and other consumer groups filed a formal petition with the PUC seeking a PG&E rate cut. The PUC, meanwhile, continued to fight PG&E about its proposal for paying its debts. Chairman Robert Glynn Jr. reported he was confident the company's plan would prevail; however, PG&E remained in bankruptcy and owed billions to independent power producers. In the year 2000, PG&E wrote off $4.3 billion in energy expenses.

PG&E and its parent PG&E Corporation went to federal court on March 22, seeking to reverse the bankruptcy court's decision on PG&E's reorganization plan that would have allowed PG&E to circumvent thirty-seven state laws. If this filing failed, PG&E could not circumvent these thirty-seven laws without first proving that the laws were primarily economically not harmful. PG&E selected the U.S. district court for its appeal rather than the bankruptcy appellate panel as it felt the panel would be biased. It contended that the bankruptcy judge misinterpreted the preemption standards, referring to a recent decision by the U.S. Supreme Court regarding a case in New York, as well as a bankruptcy PUC case in New Hampshire.

In March, Calpine said it would cancel or delay $3 billion in turbine purchases. Calpine also canceled orders for thirty-four natural gas turbines, one steam turbine, and miscellaneous other equipment. The cancellations were the result of a significant decline of electricity and natural gas prices resulting from mild weather and weak market fundamentals. Power generators were struggling with slumping electrical prices, capital spending requirements, and debt levels. Given the existing upheaval in the energy market, Calpine put on hold its 600 Mw Russell City Energy Center, scheduled for construction start in the summer. Calpine questioned its ability to obtain backing for the $300 million investment for a license approval expected in August.

April

Governor Davis was so captivated with the preceding year's 20/20 program that he wanted a replay of the program during the current year, saying, "Everything is set to go." But the idea was not popular with the PUC and some consumer groups; in the past year the program had cost

$285 million statewide, with results that were difficult to evaluate. TURN was concerned that another year of 20/20 would both cost too much and accomplish too little. PG&E estimated that in 2001 the program had cost approximately $200 for every mwh saved, in contrast to the current spot market prices of $32 to $35. Summer prices in the current year were not expected to exceed $50 per mwh. Arthur O'Donnell, editor of *California Energy Markets Newsletter*, said, "Twenty-twenty was an unqualified success in terms of raising people's consciousness, but you can turn a success into a failure pretty quickly by using it inappropriately." He went on to say that with supplies up and fears of blackouts receding, the state has no real need for power bill discounts. "I think he [Davis] is worried about November. He's trying to buy off the citizenry." David Freeman dismissed the talk, saying, "There is always an argument that funds can be better spent another way. One thing is sure, you had 20/20 last year and we achieved unbelievable results." Scott Matthews, conservation chief with the CEC, said the program was a part of a mix that drove down summer peak power by about 6,000 Mw statewide. But he said it was impossible to separate 20/20 from the impacts of higher electric rates, a statewide advertising campaign, appliance rebates, a vast array of business incentives, and sheer public spiritedness. PG&E estimated the 20/20 contribution to conservation was about 20 percent effective. CEC assistant executive director Thom Kelly believed 20/20 might have shaved off 1,100 to 2,200 Mw last summer.

The CEC predicted that the state would have enough electricity in the current year, even with an unusually hot summer, if people conserved half as much as they did in the past year. The CEC's manager of electrical analysis reported that two big plants totaling 1,800 Mw were slightly ahead of schedule, and another 500 Mw plant was expected on line by midsummer. The CEC expected to import between 1,000 to 3,000 Mw from the Northwest, the result of the heavy snowfall, along with new power plants in the Southwest. July, however, was expected to be uncomfortably close to excessive power demand, and the possibility of plant shutdowns remained a concern. Steve Larson, executive director of the CEC, said, "I don't want to encourage people not to conserve. It's absolutely essential; I think the commitment is so strong in the state not to have blackouts that we won't have blackouts." That confidence was shared by the utilities and the ISO. In 2001, the state had managed to get through the summer as a result of new plants coming on line, along with conservation and cooler weather. The ISO expected to obtain a demand decrease of 500 Mw by June from power users who would drop off the system when needed upon advance notice—for a price. The ISO kept its early-warning programs in place. The Consumers Services Department had spent about $62 million in the preceding year reminding people to conserve and planned to spend about $57 million in 2002, with most of it between the spring and fall in its "Flex Your Power" advertising program. David Freeman, chairman of the California Power Authority said, "Conservation needs to be a permanent part of our policy. By really hammering it home this summer, perhaps it starts being habit-forming."

California had been successful in renegotiating about $15 billion of its long-term contracts with two of the biggest electrical suppliers, Calpine Corporation and Constellation Energy, and two smaller suppliers, including Whitewater Energy Corporation California. State officials estimated that the State, with these negotiations, would save consumers $3.5 billion over the next twenty years. Under the Calpine Corporation and Constellation Energy agreements, the State would end its efforts to collect refunds for overcharges and would end its price-manipulation probes. The two companies agreed to pay a combined $8.5 million, thereby settling the State's claims. The power generators accepted even tougher terms, which included the building of a power plant considered important to San Diego and agreeing that the State could have the right to takeover power plants under construction if the projects were going too slow. On the producers' side, the attorney general would accept a $6 million settlement from Calpine and $2.5 million from Constellation to end the probes into possible illegal pricing practices. In addition, it was agreed that other state agencies would drop their efforts to push FERC for alleged overcharges. The PUC agreed to drop its action with FERC that was to set aside contracts for unreasonable prices. As a result of the settlement, the two generators appeared to have their major legal problems behind them.

The State estimated it could realize savings of $560 million by shortening Constellation's contracts by six months and approximately $3 billion on four existing Calpine contracts, of which $800 million would come from shortening one contract from twenty years to ten years and $2 billion would come from cutting two other contracts from ten years to eight years. Sempra Energy and Williams criticized the settlements as being too generous; however, the settlement of the contracts set a precedent for renegotiation of the remaining $28 billion worth of long-term contracts.

The PUC on April 4 voted unanimously to move the state's utilities back toward more regulation. It allowed PG&E to collect $2.9 billion for electrical energy produced in the current year; SCE was allowed $3.8 billion and SDG&E was allowed $466 million. The vote returned PG&E to "cost of service" in calculating its rates, as it had been before deregulation. The approved formula allowed PG&E to charge an estimated $716 million less in 2002 than it could have charged under PG&E's proposed plan. PG&E had proposed charging 5¢ per kWh for its nuclear and hydroelectric plants for the next twelve years, after which it would allow the charge to float with the market rates. The PUC revised the rate to 2.8¢ per kwh, a change that was welcomed by consumer advocates. The decision also (1) changed the way PG&E collected for its nuclear power; (2) modified an incentive plan to lower payments, and (3) dismissed PG&E's proposed profit-sharing plan. In addition, the decision set up special procedures for balancing accounts, allowing PG&E to track actual costs in the event that projected figures changed during the month. PG&E had been seeking more than $6 billion for its 2002 energy production and planned to give the PUC its proposed method of rate calculations for the coming years.

Legislation to allow the State to inspect power plants was passed on April 8 and would be on the governor's desk for signature by April 12. The bill would create a three-person commission charged with establishing definitive plant-operating guidelines and would require the PUC to enact and enforce the guidelines. Nuclear and municipal plants would be exempt. Dynergy, which operated three plants in the state, opposed the legislation. The PUC had been conducting plant inspections on a voluntary basis, but Commissioner Lynch commented that the PUC had met stiff resistance from some generators. She went on to say, "It is really important to be able to inspect plants when they are down. Hopefully, we can prevent some of the [power] withholding."

Attorney General Lockyer filed two antitrust suits in the U.S. district court in San Francisco, the first of several legal actions against energy marketers in which the State attempted to acquire the offending power plants. The suits called for triple damages, including return of "illegal" profits. The suits alleged that Mirant Corporation and Reliant Energy controlled significant amounts of regional power and used their clout to run up energy prices. The price run-up had started in January 2000. Both companies denied the allegations, contending the lawsuits were efforts to support Lockyer's reelection campaign.

It had been generally observed that California's new power plant construction activity had begun to lag, a trend that began the preceding year. The causes for the slowdowns were attributed to money. Developers had been trimming spending—watching their bottom line. An additional factor credited California's unsettled energy markets. The long-term effect was unclear, and there were some concerns that serious problems could surface after 2005. Seven of the sixteen plants approved for construction had been put on hold for up to two years. Fourteen of sixteen projects that sought licenses notified the CEC that their construction schedules would not be met. Additionally, in April the CEC was notified that two large plants were being delayed. The two plants in question would have added 1,500 Mw to the grid. Senator Steve Peace (D-San Diego) introduced SB 28x for consideration. The bill would have toughened deadlines for plant construction. Peace's aide, John Rosza, said "There was concern that power plant delays might leave us in a vulnerable position in three to four years. We want to make certain that when somebody gets a permit to build a power plant and ties up resources to be able to build it—and that includes location, air permits, water rights, all sort of things—that the power plant gets built." Permission to build a power plant was a valuable asset and once obtained, the license could be sold for a profit. There was no concern at that time that it would be necessary for the State to step in to build plants, as 2,500 Mw of new power had come on line in the past year, and 2,900 Mw were expected to be brought on line in the current year. The ISO had projected 1,000 to 1,500 Mw were required to be added to the grid each year. In addition, the ISO stated that California was likely to lose 1,000 Mw of aging plants in the next few years due to aging or the reluctance to upgrade plants to meet air standards. The CEC planned to do an analysis of the power plant construction slowdowns.

May

The DWR renegotiated a long-term contract with Calpeak, a company based in San Diego. The Calpeak contract was included in the PUC's filing with the FERC in February that sought to have the long-term contracts either voided or modified. Originally, Calpeak was required to build seven peaking power plants and sell its power for ten years. The new deal reduced the number of plants to six. The elimination of one plant saved the State $58 million in capacity payments. Of the remaining six plants, five had their capacity payments reduced, resulting in a savings of $7.5 million. The construction of the sixth plant was postponed for eighteen months. One of the five plants was to be relocated to an undetermined location in Northern California to ease capacity requirements for Path 15.

The ISO had projected a surplus 2,000 Mw of power would be available for the coming summer. On May 1, the ISO sought new authority from the FERC to control California's electrical system. It requested (1) an ability to encourage long-term power sales; (2) authorization to impose wholesale price caps (suggested at $108 per mwh); and (3) provisions to lock up electrical supplies one day ahead instead of relying on the daily spot market. It provided price incentives to encourage construction of new power plants closer to load centers.

Calpine reported on May 2 a quarterly loss of $74.3 million, or 25¢ per share, compared to last year's profit of $119.7 million, or 36¢ a share. The loss was attributed to the disallowance of a $168.5 million pretax charge for cancellation of equipment orders.

On May 7, the FERC reported it was going to probe more intensely, looking for more evidence of price manipulation in California's energy markets. Cited was the release of Enron's memos detailing schemes for driving up energy prices. The FERC also said it would soon direct all sellers of California energy to produce internal documents regarding the use of Enron schemes or similar schemes, and it cautioned the generating firms not to destroy any of their documents. Those released memos revealed the details of how Enron traded power in a number of ways to drive up wholesale prices. California's U.S. senators and Governor Davis said Enron's conduct underscored the need for FERC to renew the wholesale price cap scheduled to terminate September 30. The senators also called for criminal investigation of Enron by the Justice Department. They felt that FERC should approve refunds to the State for fraudulent overcharges and that the long-term contracts should be renegotiated. Davis believed Enron's mischief strengthened California's case in asking for $9 billion in refunds.

May 8, FERC ordered 150 energy companies to review their records for any sign of market manipulation similar to Enron's mischief. They were to respond within two weeks of the notification with sworn affidavits signed by a top corporate officer or risk losing their license to sell power. The agency said, "Your company's failure to respond in a timely and complete fashion may result in enforcement action, including the issuance of an order directing a public utility to show cause why the commission should not revoke the public utility's authority to sell whole-sale electricity and/or ancillary services at market based rates." Corporate officers were admonished to thoroughly search company records and based on that search, declare or deny whether or not their company had participated in any of the twelve detailed schemes.

FERC's notice, posted on its web site, described the schemes outlined in two of Enron's memos from December 2000. For example, "Inc load" traders gave the ISO one day in advance a list indicating their customers' demand, which matched the energy the traders wanted to transmit through the grid. The trader would turn in "dummy" loads, which would disappear when the hour arrived, and which left a surplus of energy that was sold when prices were higher. Another scheme involved taking advantage of financial incentives of clogs on transmission facilities. Phony clogs on transmission lines could be created by scheduling energy transfers in advance and then collecting payments for not sending the energy. Another had to do with "get short/sell short[13]," a specialized kind of energy trade used to strengthen the grid that required telling the ISO exactly where the energy was coming from. It involved paper trades with no ultimate supplier. The memo said it was necessary to submit false information that purported to identify the energy source. Yet another had to do with mislabeling energy—the company would purport to sell dependable "firm" power that included extra features to make it dependable and then deliver "non-firm" energy that lacked those backup assurances. Then there was megawatt laundering, an apparently legal ploy to get around government price caps. It involved buying power in the price-controlled markets and selling it outside the state to a buyer who, for a small fee, agreed to sell it back. Then when the critical instant came to avoid blackouts, the trader would resell the energy back in real time for excessive prices because the emergency allowed the price caps to be exceeded. There were times when traders could legally buy energy in California at $250 per mwh and sell the energy for export at $1,200. This, of course, may have led to declarations of Stage 2 emergencies within the state.

FERC observed that the strategies intended to maximize profits by taking advantage of California's energy crisis. The tactics were not only

---

13 Selling energy that didn't exist.

used by Enron but by other traders as well. If FERC found any evidence of participation in any of the twelve schemes, the companies were directed to describe each event in which it had engaged in such activity, and they were to provide full disclosure of the details and any related documents. The companies were also to state whether they engaged in schedule manipulation to take advantage of California's transmission congestion, a practice Enron said was typically practiced by other traders. They were also asked whether they had engaged in buying and/or selling power through out-of-state marketers so as to bid up the California wholesale price.

Enron's memo could not have come at a better time. The State was involved in a controversy with the FERC for refunds based on price gouging. Davis had demanded for the past year that FERC order refunds of $8.9 billion. Before the exposure of the memos, the State may have been limited in its recovery because of commission rulings; however, California and the Northwest now felt they had a firmer legal case to get larger refunds. David Freeman said, "We ask for only $9 billion; we are probably entitled to close to $30 billion." California's case entered a crucial phase in July, wherein the State could very possibly force the commission to expand the period for which the State would be entitled to refunds. Mark Shultz, vice president of public affairs for the Electric Supply Association, said his group didn't believe firm evidence could be gleamed from trading memos obtained thus far. He went on to question whether Enron had indeed "gamed" the market and suggested, "There should be no rush to judgment." Pressure on the commission may have become more intense after U.S. Senator Feinstein released the memo written by SCE economist Gary A. Stern. The memo, in August of 2000, revealed to the commission the schemes that were used by generators to game the market. That was four months before the Enron memos were written. A decision in California's filing wasn't likely to occur until late in the current year. However, Stern stated that the memos could result in expansion of the case with the possibility of additional billions of dollars for the State. To date, the FERC had made two critical decisions in the refund proceedings: (1) limiting the time in which the State would be entitled to refunds to between October 2, 2000 and June 30, 2001, and (2) declaring that power purchased by the DWR would not qualify for refunds.

Stern believed those decisions reduced the State's potential recovery to about $1.5 billion, which was $7 billion below what Davis had been seeking. Approximately half of the reduced amount was due to FERC's prevention of recovery prior to October 2, 2000. FERC had concluded there was no valid proof of violations prior to that date. Under the Federal Power Act, the commission could not consider refunds until sixty days after of the filing of a complaint that challenged wholesale rates—SDG&E had filed their complaint on August 2, 2000. Stern argued the Enron memos proved that conclusion to be wrong and went on to say, "It could certainly be argued that the Enron memos are the hardest evidence yet that tariff violations were going on prior to October 2, this could give the FERC a basis for considering refunds before October 2." A FERC judge was expected to reach a decision on what the acceptable price for wholesale power was during the crisis.

On May 15, Senator Feinstein introduced to the commission a memo alerting the commission to deceptive practices, which were almost identical to those contained in Enron's memo of August 2000. The memo was written by Gary A. Stern, an economist and director of market monitoring and analysis in the Regulatory Policy and Affairs Department of Southern California Edison. The purpose of the memo which was to satisfy FERC's inquiry and to initiate a FERC investigation. The memo caught FERC chairman Pat Wood off guard because it implied FERC was aware of the abuses and had done nothing to stop them. The memo became the basis for an internal report that resulted in a December 15, 2000 order. Wood said, "The report became the template for the December 15 order that got rid of a lot of the flaws. I don't know if everything in that memo got fixed in that order." The question then became whether the memo was sufficient evidence for the FERC to void those contracts, thus avoiding the outrageous energy prices contained in the costly long-term contracts that had been entered into by California and the Pacific Northwest.

On May 22, FERC was inundated with documents in response to its order for the reporting of any marketing mischief; the order had been directed at 150 energy marketers. Duke, Reliant, Dynergy, Mirant and Williams acknowledged they may have participated in some of the features detailed in Enron's memos, but all underscored they had followed the rules and had done nothing wrong. Williams's president said, "We closely monitored the evolving market and made every effort to participate in a way that was fair and legal. Williams does not have, and has never had, strategies to engage in illegal or improper behavior." That was pretty much the tone that SMUD experienced in conversations with Enron regarding ways that Enron could profit by dealing with SMUD. SMUD didn't accept the proposal. Former ISO chief operating officer Kellan Fluckiger said, "My guess is that everybody is going to say they didn't break the rules." He also commented, "We have a structure that is inherently so flawed that the number and scope of these opportunities is staggering. The rules are so loose that nothing is specifically outlawed, and there are few consequences for misbehavior." Economists commented that these had been little more than a detailing of market tricks that were widely known or suspected during the power crisis. There was a host of others, including Calpine, LADWP, and PG&E, which had publicly announced that they did not participate in Enron's tactics. The five companies that had purchased the state's power plants acknowledged some of their practices to the FERC. Dynergy, Reliant, and others described some misrepresentation of projected demand but characterized it as a routine procedure for testing purposes. Mirant acknowledged that in one instance it had participated in megawatt laundering by selling power out of state and repurchasing it to avoid the wholesale price cap. Mirant had declined to make its FERC filing public. Williams acknowledged it was paid to relieve congestion and was unable to rule out megawatt laundering or exporting for profit. SMUD, in its filing, reported two irregular incidents that included trading offers from Enron and AES. AES, an Ohio-based trader, had shipped power to SMUD. The Ohio firm said the shipment was a mistake and corrected it at a loss, so it denied doing anything improper. Fluckiger said FERC would find many discrepancies when it reviewed the filings. "If [FERC officials] are honest, it will lead to the conclusion that far larger refunds and adjustments are due than have been proposed by FERC up to now."

June

FERC finally conceded California's unrestrained energy market really needed attention. Pat Wood, FERC chairman, decided to dedicate up to one hundred people, a tenth of its budget, to a new office charged with oversight of the daily activity of the power markets nationwide. This new Office of Market Oversight and Investigations would be responsible for identifying market manipulations and tracking daily energy markets nationwide and providing a safety net for states with deregulated markets, including California. A Washington DC energy attorney, who had dealt with FERC for about two years, said, "FERC is really on the hot seat now; they obviously need to have this in place very quickly." The agency had been under severe pressure from the public, the Bush administration, and Congress to get off the dime and address the allegations, including the increasing turmoil that had been pounding the energy industry. The Enron memos helped bring the entire energy market problem into focus. FERC commissioner William Massey, who had been championing California's case, said, "The average American looks at this and thinks if that's what electricity markets are all about, we want nothing to do with them, whatsoever." He went on, "Job one, right now, is to stop the funny business. Job two is to take the necessary steps to restore confidence." Loretta Lynch said, "I don't have any faith in FERC to be able to . . . ensure just and reasonable rates. Why should we believe the rhetoric again?" Pat Wood acknowledged that it was reasonable for Californians to be skeptical given the blackouts and price spikes and the repeated clashes between state officials and FERC officials. California had been seeking price caps, long-term contract renegotiations, or contract cancellations in the amounts of billions of dollars in revenues. FERC would now be under the public's scrutiny in its regulation of electrical and natural gas markets, oil and gas pipelines, and hydroelectric licensing. Gary A. Stern of SCE indicated he was reassured that they were treating the problem with more seriousness and that Pat Wood was trying to do what he had said he would do, and that was if any wrongdoing was found, he would come down with a heavy hammer.

The General Accounting Office, after a yearlong investigation, released a report that took FERC to task, stating, "FERC does not currently have enough staff with skills and knowledge of competitive energy markets to effectively regulate and oversee these industries." These were conclusions that had been reached by California's congressional delegation, Governor Davis, and utility officials. U.S. Senator Joe Lieberman (D-CT), who headed the Senate Governmental Affairs Committee, called the report's findings a "validation" of California's criticism that FERC had done too little, had been too late in responding to skyrocketing wholesale rates, power outages, and rolling blackouts before the crisis finally eased last spring. He stated, "The case is made here that FERC was aggressive about deregulation, not aggressive about—in fact, extremely passive and weak about—oversight, and the result is that some very slick game players took advantage of that opening, and a lot of consumers, particularly in the

West, suffered." The auditors said, "Absent an effective regulatory and oversight approach, FERC lacks assurances that today's energy markets are producing interstate wholesale natural gas and electricity prices that are just and reasonable." The report pointed its finger at Congress for not giving the agency the tools to combat wayward merchants. "Without a meaningful range of penalties, FERC lacks adequate enforcement 'bite' to deter anti-competitive behavior or other violations of market rules." FERC chairman Pat Wood attached a response to the report in which he agreed his agency had failed to meet its obligations to protect consumers. "The commission's internal restructuring to support its new market oversight role has not kept pace with the speed of energy industry restructuring. Today's staff is better suited to regulate cost-of-service rates than market-based regulation." Wood stressed he hoped the new office of market oversight would probe complaints and pursue enforcement actions when unfair trading occurred and would catch up with oversight of competitive markets. Lieberman said he would ask the GAO to monitor "on almost a daily basis" how well the new office functioned. "He says clearly that this agency didn't do what it was supposed to do, and now we're doing it," Lieberman declared. "I think we've got give him [Wood] a chance on this one to prove that that's in fact what he is capable of doing." U.S. Senators Barbara Boxer (D-CA) and Dianne Feinstein said the FERC issue was reason enough to extend the West coast price caps.

State regulators had been charging El Paso Merchant Energy of doing "wash trades" in California markets. El Paso Merchant Energy responded with a qualified denial to FERC regarding charges that it had done "wash trades." They denied selling natural gas back and forth at the same price on the same day. El Paso claimed they depended on the textbook definition of wash trades where the trades were used to deceive other traders but said they never used the method. This was not the definition used by FERC. Reliant had acknowledged it used wash trades as a method of making the company appear larger than it actually was, contending it was a way of making Reliant appear larger in order to attract more business. Observers from the industry said that might have inflated prices, thus hurting the consumer. El Paso and related companies were accused by the PUC of using their pipelines as well as their close relationships to boost natural gas prices. Reliant said while it had done no wash trades in the West, it did admit some superficial trades might have resembled wash trades, stating these were coincidences which were intended to reduce risks or to correct mistakes.

The state attorney general had filed a suit with the bankruptcy judge against PG&E Corporation for $4 billion for the benefit of the ratepayers. The suit contended it was money that had been transferred to the corporation before PG&E had declared bankruptcy in 2001. PG&E wanted the case tried in the bankruptcy court on the basis that only the utility was legally allowed to request a refund. The bankruptcy judge ruled that the public also had the right to request a refund declaring that key portions of the case fell under the state antifraud law and not under the federal bankruptcy code. The judge said the suit properly belonged in the state superior court; however, not all of the state's claims belonged there. The court mandated several claims in the suit must be tried in the bankruptcy court. Included were those claims that were related to the legality of the corporation's actions after the utility filed for bankruptcy, including those claims that related to circumventing the state's laws and regulations.

Lawmakers (mostly Democrats) called on Davis to end the state of emergency that he had declared on January 17, following rolling blackouts. In addition the National Tax Limitation Committee was threatening to sue for the end of the emergency. However, many emergencies were purposely kept open to assure continuation of federal disaster-relief funds. Up until now Davis had called upon his emergency powers twenty-three times, and among those actions he had expedited power plant approval and had seized the PX's long-term power contracts. The state Emergency Services Act allows the governor to declare an emergency and invoke, as necessary, a broad range of emergency powers. The argument for the end of his emergency powers was that the balance of power between the branches of government had been upset and that continuation of such emergency powers constituted a violation of law. Senator Bowen said, "Emergency powers temporarily upset a constitutional balance of power between the governor's office and the legislature, which is the house that more closely represents [the] citizens. An emergency declaration basically makes the governor the supreme dictator while it's operative." However, Bowen acknowledged if there was still a problem, the governor could always declare another emergency, if the situation warranted it. Those that were in favor of the governor retaining his temporary emergency powers argued that it was unknown what was going to happen in September when the federal price caps were scheduled to end. In the past year, Davis had used his emergency powers only twice. On July 30, 2001, he suspended a law that would have allowed PG&E to transfer its Kern power plant to the North American Power Group. That would have allowed the North American Power Group to sell power to the State. On May 23, 2002, he ordered the DWR to implement the 20/20 program for the summer. Bowen's resolution, terminating the temporary powers, had passed the senate the previous September. However, the assembly had yet to take any action on the resolution, and it was unlikely to reach the assembly floor. The support for Bowen's resolution came from her fellow Democrats; the legislature Republicans opposed it.

A state senate hearing on June 27 revealed that the LADWP may have participated in some schemes similar to those revealed in the Enron memos, particularly the scheme known as "Ricochet." Transcripts and an email conversation dated November 12, 2000, were the bases for the suspicion. Ricochet was a technique where power was sold for export out of state and re-imported at a higher price. "The net effect of this transaction was to take megawatts sold in Southern California at $50 to $75 per mwh and . . . turn those into megawatts which were returned to California at what we believe was $250 per mwh," according to Larry Drivon, attorney for the special senate committee. An employee at PG&E's trading division had warned, "Beware of ricochets [with LADWP]". The senators focused primarily on transcripts of conversations among LADWP energy traders, the ISO, Constellation, and the PG&E National Energy Group. There was one instance where a LADWP trader scolded another trader for revealing too much information about the transactions with the ISO. LADWP officials denied any participation in Enron's ricochet transactions. LADWP's manager of wholesale trading claimed the transactions differed from ricochet because the power never left the state: "The LADWP did not engage in any trading of power across state lines to avoid price caps, [as] that transaction would be characterized as a 'bounce-back' . . . The November 11, 2000, transaction was completed consistent with LA's good neighbor policy of playing by the rules." The committee members did not accept the utilities explanation since the net result had been that of forcing up of power costs.

Path 15 continued to be a bone of contention after January 2001; the addition of a third line was considered a must resulting from a blackout that occurred because the Path had insufficient thermal capacity to transport the required electrical demand to Northern California. Now that the dispute had cooled off, even the ISO admitted that the additional line was not needed for basic grid dependability. However, the ISO did believe the additional line would remove a bottleneck the power predators had been able to exploit. The estimated $306 million for the Path 15 expansion was to be paid for by millions of California ratepayers. The question was raised, was this another cost of the electrical deregulation? Private builders were seeking higher rates of return and shorter depreciation time than what had been allowed under regulation. The grid operators said Path 15 didn't cause the problems but only impeded the ability to transmit power from Southern California to Northern California at a critical time. The PUC had ordered PG&E to look into the expansion of the Path, while President Bush ordered the WAPA to help fix Path 15 as well. FERC approved an unusually rich repayment formula for Trans-Elect, a private company that would be working with WAPA and PG&E to add the third high voltage transmission line. WAPA hoped to commence construction in the spring of 2003, with completion in 2004. The PUC, at the time, was still studying whether the line was really needed. Apparently, PG&E had looked into the feasibility of adding a third line twice before but couldn't justify the costs. Since mid-2001, three of four large power plants had come on line in Northern California but only one in Southern California. That eased the need for the additional transmission line, but Jeanne Sole, the ISO's regulatory counsel, said the bottleneck still made it easier for marketers to jack up prices when one region was isolated from the other. The consumers' cost for the additional line was estimated to be about $60 million a year for at least ten years, according to the PUC's Office of Ratepayers Advocates. That additional cost was calculated to be about on .3¢ per kwh. The money would be collected by the utilities that used the ISO-managed transmission lines. PG&E would make payments to the ISO who, in turn, would pay the builders at rates set by FERC. The PUC had contended FERC payments were unusually high.

July

July 2, DWR filed suit against Sempra Energy Resources, claiming that Sempra had not fulfilled its promise to build a 570 Mw power plant in Bakersfield by April 2002. As a result, the DWR wanted out of its $6.6 billion, ten year contract. The DWR had claimed that Sempra was operating in bad faith by not building the plant and was purchasing energy on the open market and reselling it to the State. Sempra argued the contract did not require any power plants to be built; it only required that a given amount of power be delivered. However, Sempra did say that a plant was being built and that it would be on line next summer.

July 11, FERC increased the wholesale energy spot price in the western states. The rationale was to ensure sufficient supplies to meet demand during summer hot spells. FERC was worried that the ISO's ceiling price had been set too low and would discourage generators from selling in California, even though they were required to sell if they had spare power. The action negated the policy that allowed the ISO to use various factors, including the price of natural gas, in calculating the maximum price. The order said, "We act now because we cannot expose customers in California and other western states to the risks of a low price cap. Returning the price cap to $91.87 per mwh should continue to facilitate energy offers into the western market." This fixed maximum price would remain in place until September 30, 2002. The approved change factored in the price of natural gas. At the time, California's power requirements were assured by the DWR's long-term contracts and were not affected by the price cap. Only power demands in excess of the amount of power supplied under the long-term contracts would be affected. Governor Davis said, "There was no need for FERC to act. The DWR was able to buy all the power needed in the $20 to $30 per mwh range. If FERC believes nearly $92 per mwh is a just and reasonable rate, it is out of line. Once again, FERC is protecting generators, not consumers." FERC, at its scheduled meeting in the following week, planned to resolve competing pleas by the State, generators, and others to further refine the pricing mechanism set to expire September 30.

On July 17, PG&E's 74,000 creditors were to be mailed ballots inviting a yea or nay vote needed by August 9 on the competing reorganization plans to resolve the company's bankruptcy case. The competing proposals were written by PG&E and by the PUC. The official creditors committee, which originally favored the PG&E plan, was recommending an affirmative vote on either plan as they lacked any preference on either plan at that time. The creditors' chief concern was the legal activity PG&E faced in getting their plan through both the courts and the federal regulatory agencies. On the other side of the coin, the PUC plan did not guarantee PG&E a stable flow of sufficient rate revenues. In all Chapter 11 Bankruptcy cases, balloting is structured into certain standard classes. For the

PG&E plan, there were eleven classes, with eight classes for the PUC plan. If yes votes from at least half of the creditors were returned within the class (of those holding at least two thirds of the debt), that class would have been approved. PG&E Corporation, the parent company, had the only vote in the class of "common stock equity interests." However, the parent company held all of PG&E (the utility) common stock. The PUC plan had no corresponding class. If just one class accepted a plan, the bankruptcy judge could force that vote on all classes by making a judicial determination that it's fair to them. Hearings were set for the fall at which time the judge would decide whether the plans were both feasible and legal. Each plan had its hurdles to overcome. PG&E had to convince the judge that circumvention of state rate and environmental regulations was an economic necessity, while the PUC had legal issues to resolve, including a PG&E threatened suit to prevent dilution of the corporation's stock The FERC believed aggressive federal action was needed in order to bring the nation's electrical systems under control. Such federal action, they believed, would avoid a replay of the California crisis.

In its meeting of July 17, FERC, by a vote of four to zero, raised the cap on wholesale electricity, to be effective October 1, contending that the increase was needed to ensure flow of power into the state. The new cap was increased from $91.87 per mwh to $250 per mwh. In addition, it ordered the Davis appointed ISO board to be replaced by a new "sufficiently independent" board by January 1, the members to be named later. In June 2001, FERC had ordered a price cap that floated with generating costs and other factors. Such a price cap was considered necessary in order to prevent megawatt laundering and other dubious market strategies. Critics claimed that the decision to set the higher price cap would lead to inflated prices for spot market purchases. FERC contended that the new higher cap would encourage the construction of new power plants, transmission lines, and natural gas pipelines. According to FERC, the new cap was "striking a balance between the need for investor confidence in the markets to bring necessary generation and infrastructure on line and the need to protect against market abuse." The order also maintained the requirement that generators be required to offer all of their power for sale to prevent the withholding of power to drive up prices. The power sellers had asked for a cap at $1,000. The current California spot market prices had been running around $35 to $40 per mwh. Gary Ackerman of WPTF said, "People are delaying and canceling plants left and right. People can't do business here." He also commented that the existing price ceiling was discouraging new plant construction. Arthur O'Donnell, editor of the California Energy Markets newsletter, opined that lower price caps could sometimes inhibit power generation. Davis called FERC's decision a sellout to power generators, "There's no justification for tripling the amount of money that generators can siphon out of the pockets of California consumers. This order just invites more mischief by energy generators." Doug Heller, an FTCR consumer advocate, believed FERC's decision was misguided, particularly in light of disclosures of alleged market manipulation by Enron. "This will allow the power companies to go back to the profiteering behavior that . . . created economic disaster."

On July 31, FERC proposed a plan to standardize electricity trading and transmission access across the contiguous continental United States. The FERC approved a draft of a proposal that would place the FERC as the major player with increased control of pricing, planning, and operational decisions that had once belonged to the individual states. That market proposal prompted considerable static from fifteen states; they warned that approval of the plan would force the nation toward deregulation, thereby driving up prices in those states where electrical energy was still inexpensive. FERC argued that the plan represented a solution to the troubles that plagued California and at the same time was the way to improve electrical markets overall. The plan envisioned a national grid, overseen by independent regional boards that would be in a position to determine when and where new power plants needed to be built, including the transmission facilities to service them. The transmission network would be available to all generators under a single rate structure set by the commission. There would be new rules, yet to be designed, to prevent the kinds of tactics that allegedly helped drive up prices in California in 2000 and 2001. The rules would ensure adequate power would be always available to all consumers. The new rules would encourage long-term contracts, thus minimizing potential price swings on the spot market. There would also be regional transmission price differences designed to avoid transmission bottlenecks and congestion. The new rates would additionally serve as incentives for investors to build new facilities. The impact of the plan was expected to be the greatest in those states that were still regulated; it would require them to surrender control over their transmission systems to the FERC's regional agencies. FERC said they wanted to respect the needs of municipal utilities, but the full impact of FERC's proposal could not be determined until a full six hundred-page document was released. California's ISO was determined to closely review the full document in order to fully evaluate how the plan's rules would deal with potential market abuses. The Consumers Union welcomed the idea of a standard market but had its reservations, since FERC had serious credibility problems due to its past failures. The FERC would accept comments for the subsequent two-and-a-half months and targeted September 30, 2004, as the date for implementation. FERC's plan had seventy-five unresolved issues on which it had invited comment; however, the basic plan outline remained unchanged from past proposals.

August

The California state ISO voted unanimously on August 7 to contest FERC's order to replace its board with a board free of State control. This put the ISO on a collision course with FERC. There was concern that the issue could eventually go before the U.S. Supreme Court. The question was, who had the right to control the ISO?

FERC's biggest contention was that there was a conflict of interest with the DWR buying electricity for the state's utilities and the ISO running the power-buying system. The head of the Independent Energy Producers Association, Jan Smutney-Jones, urged the FERC and the ISO to work toward compromise before regulatory uncertainty stymied power plant construction more than it already had, saying, "We're one bad snow pack away from repeating the crisis." And he noted that a court fight could last for years. He called on the ISO to urge FERC to negotiate with the State to develop a government structure that would make sense for everyone. There was a real question as to who was in charge, since the ISO was a state-created entity regulated by FERC. The state legislature had passed a law in January 2001 that gave the governor the authority to appoint the board. FERC pointed out that that law was passed after the FERC had already ordered a restructuring of the board in order to be free of State influence.

August 13, FERC released a report that the commission characterized as being tough on market manipulators; the report said natural gas prices used to calculate refunds might have been manipulated. FERC launched formal investigations and held hearings into the behavior of Enron subsidiaries: Enron Power Marketing, Enron Capital and Trade Resources, and Portland General Electric Company, along with Avista Corporation of Spokane, Washington and El Paso Electric Company of Texas. All were suspected of manipulation of the electrical prices, and FERC could consider targeting their corporate profits for refunds. FERC once again asked Congress for more authority to punish abusers. Governor Davis called the report a "whitewash," while consumer advocates and some within the Davis administration greeted it optimistically. Mike Florio said, "I am certainly surprised. If they really carry through on what they're suggesting, it's one of the biggest breakthroughs in this whole sad story." Energy marketers, fearing the prospects of larger refunds to California, thought the investigations might hamper ongoing efforts to renegotiate their long-term DWR contracts. "I don't think this works for the sellers' advantage. The industry is going to take this very seriously," said Gary Ackerman, head of WPTF. If the companies were found to have violated federal law, their ability to trade power could have been revoked and they could be ordered to make payments from their profits. One promising plus was the fact that the commission staff considered recalculating the formula that was used to award refunds to western power purchasers. FERC had been using published prices of natural gas to calculate how much electricity should cost. FERC staffers said the formula should be junked, as sellers and traders could have easily misled industry publications. The report said, "Unscrupulous traders could manipulate natural gas price indexes in order to increase profitability of their electricity positions." FERC issued a request that asked for comments within the next thirty days on a proposed alternate pricing formula, which would be based on what natural gas sold for at the production field, plus transportation costs. FERC had repeatedly hesitated to pursue buyer's demands for refunds through the exclusion of time frames and types of trades. The report said, "Enron, as a corporate entity, displayed great eagerness to experiment with all aspects of the market rules and protocols in an effort to 'game the system' or to provide false information." Governor Davis said, "FERC is anemic. FERC is spineless." Upon hearing Davis's remarks, FERC commissioner Nora Brownell said that while FERC is determined to halt market manipulation, "We do have due process. We do need to actually prove it."

The ISO had on August 13 asked the FERC to stay its suit in order to allow time for all parties to reconsider their positions. The California attorney general stated that by complying with FERC's order they would violate state law. Three days after the request, FERC proceeded with its suit. Roger Smith, a Washington DC attorney specializing in energy issues, observed that both parties had long held opposing positions. He said, "Now it's clear to FERC, and probably to California, that what [is] needed to move this issue off the stalemate is a court decision, and the sooner the better, I guess."

Davis wrote the ISO on August 16 saying, "FERC continues to press its failed deregulation agenda." Davis appeared at the ISO's board meeting the following day, arguing that the order for a new board was "nothing less than a hostile takeover of California's electric grid by the federal government." In its vote the board elected to fight the FERC order in three ways: (1) ask FERC to reconsider its July 17 order calling for a new board, (2) request FERC to stay its order during the reconsideration period, and (3) ignore the FERC order to immediately begin the restructuring of the ISO board

August 16, FERC filed a suit in the U.S. district court in Washington DC that was intended to force California's ISO to create a new ISO governing board. The move startled both FERC watchers and grid operators. The filing prompted an immediate response from Governor Davis, whose spokesman said, "We'll come out swinging [and will use] every tool at our disposal." In its suit, FERC contended that the governor-appointed board appeared to be biased in favor of California's customers and was hampering efforts to create regional transmission organization. The suit also contended that the ISO was in violation of the Federal Power Act that gave FERC sole authority over all ISOs. State officials argued that they wanted to retain control in order to ensure that the power generators never again would gouge California's consumers.

On August 16, U.S. district judge Vaughn Walker reserved his decision on PG&E's appeal of the bankruptcy's judge ruling that denied PG&E's right to disregard a group of state laws and regulations. Walker observed that Congress had amended the bankruptcy reorganization process, taking the states out of the process. Walker said the bankruptcy ruling "stands alone" among judicial decisions on the State's right to interfere with a debtor's reorganization. Lawrence Tribe, who argued PG&E's case, said that a reversal by Walker would eliminate the bankruptcy judge's reason for considering rejection of PG&E's reorganizing plan: "Every court has it right, and the bankruptcy court in this case and the PUC have it wrong." He went on to say PG&E was only seeking the right to override laws that stand in the way of its reorganization plan and only during a transition period. Walter Rieman, representing the PUC, told Judge Walker that "PG&E has continually said preemption was an across the board, 'take no prisoners' position that would sweep away state and federal regulations." Karl Fingerhood, attorney with the U.S. Justice Department, argued nothing would stop PG&E from displacing federal laws that protect the air and water and the control of nuclear handling regulations. Walker responded that any legal violations would only be temporary.

On August 22, the creditors committee became a cosponsor of the PUC's reorganization plan involved in PG&E's bankruptcy. The plan would remove the possibility of PG&E's preemption of state laws and regulations and if accepted would remove the case from the federal court. The committee had been supporting the PG&E plan but now became concerned about the possibility of PG&E's plan being tied up in the courts due to court challenges on constitutional grounds. Lynn PoPucki, a UCLA bankruptcy law professor, said that the committee's position should persuade rank-and-file creditors to get behind the PUC plan. In his opinion, that support would have been persuasive with the bankruptcy judge. He stated, "Creditors virtually always vote in accordance with the recommendations of the creditors committee." A breakthrough occurred after PUC gave assurances that securities used to finance the plan, including $1.75 billion in new preferred PG&E stock, would be backed by sufficiently high retail rates. Consumer advocate Harvey Rosenfield felt that the agreement was an "outrageous and illegal" deal, saying it would lock in rate increases that should be returned to the ratepayers. The PUC counsel replied that in their opinion, the agreement should have no influence on an anticipated 2003 rate increase.

U.S. district judge Vaughn Walker ruled on August 30 that PG&E could skirt state laws concerning retail power rates as well as public health and environmental regulations, thereby allowing PG&E to move forward with its bankruptcy case. This judgment overturned the earlier bankruptcy court ruling that would have forced PG&E to justify each state legal provision it planned to skirt. Walker cited "powerful evidence" in the federal bankruptcy statute that stated that Congress intended "that public utilities should no longer be subject to costs, delays, and uncertainty" of the state approval process when the utilities seek to reorganize their debt. Lynn PoPucki, the UCLA bankruptcy professor, predicted that if the PG&E plan was approved it would provide incentive for other utilities to follow the same route to reorganize their debt. The PUC immediately predicted rate increases. State laws bared any transfer of PG&E entities until 2006; however, Walker's decision would preempt it. Gary Cohen, PUC's general counsel, stated, "If the PG&E plan is confirmed, it would certainly mean a rate increase for Northern California's families and businesses." It would allow the utility "to sell or permit development on thousands of acres of priceless watershed land, which would be illegal under state law if not for the bankruptcy plan." PG&E reemphasized that it "seeks to preempt only a limited number of state laws and regulations necessary to allow it to emerge from bankruptcy." After reorganization, PG&E and the spun-off companies would again be subjected to "all applicable federal, state and local laws, including environmental, public health, and safety regulations." Walker, in his opinion, remarked that it was "highly unlikely" that the bankruptcy judge would seriously consider confirming a plan based on "ongoing illegality." If legal violations did occur after reorganization, the opinion said, the appropriate state and federal laws would apply.

The CEC received a letter from FPL Energy withdrawing its application for licensing of its proposed 500 Mw power plant in Rio Linda/Elverta. FPL spokesperson said, "Money is tight, and companies are going to have to be very much more selective on which plants to take forward." In its letter, FPL cited "an unfavorable electric market" as one of three factors involved in withdrawing its application. FPL was also dismayed by the time-consuming efforts to buy adequate special credits to offset the anticipated plant's pollution and by "significant community opposition." Experts pointed to accounting scandals and tumbling stock prices for investors' reluctance to lend money for the construction of new power plants. Also, it was observed that the issue between California's ISO and FERC worried investors. In addition, FERC's proposed nationwide market rules were being opposed by many states. There was also the question of who would buy power for the utilities after January, when the DWR's purchasing authority terminated. The CEC's chief deputy director opined that the state's overall power outlook had not worsened: "We have a pretty good knowledge about the pullbacks for the next couple of years." He said the state had enough electricity to meet its needs through 2004 and possibly 2005.

Enron asked the CEC to suspend its application to license its proposed 900 Mw power plant known as Roseville Energy Facility in Roseville. They cited the lack of funds and its inability to find a buyer for the project. The request was part of a general trend to place power plant construction on hold. Enron had tried for eight months to obtain funding to pursue the Roseville project to completion. It was Enron's intention to continue the licensing process once funding had been obtained. The CEC was expected to grant the suspension. Roseville didn't want the project scrapped because of the monetary payments that the project would produce for the city. For example, Enron had paid the city $600,000 to date and the next payment of $100,000 was due the following October. Upon completion, Enron would be required to pay the city $2 million a year for twenty-five years. For the past year, there was uncertainty of funding, and there was the bigger question of where was California going in the future energy markets. Both factors had impeded existing construction as well as applications for new plants. The CEC disclosed that requests for suspension of power plant projects had become fairly common.

September

A new bill, SB 1078, designed to take California "to the head of the class" for renewable energy, allocated $70 million annually for renewable energy. It was a bill Davis vowed to sign into law. The concept, called a "renewable portfolio standard," was a simple one and would require all private utilities and other energy suppliers to provide 20 percent of their energy from renewable sources by the year 2010. A number of other states had more modest renewable energy standards in the range of 5 percent to 10 percent. A major energy bill before Congress called for a fixed standard of 10 percent. Supporters of renewable energy argued such a standard would help wean the nation off of fossil fuels. Environmental advocates believed the standard had a financial benefit in that renewable energy was not subjected the price fluctuations of fossil fuels. The target of 20 percent was eventually dropped in favor of an annual rate of 1 percent with a final target of 17 percent. A complex funding formula was created that would require the PUC and the CEC to cooperate, two agencies not known for working together very effectively. Builders of renewable generation facilities would be restricted from tapping the pool of subsidy funds unless they paid the "prevailing labor wage," which adds to the construction cost. Additionally, a new approval process could lengthen construction schedules. Opponents of the bill were in agreement that the additional various benchmarks and financial requirements included in the bill would have made it conceivable that the targets could never be achieved—Jan Smutney-Jones, head of IPPA, said, "It [the standard] will not work particularly well; it was motivated by symbolism . . . We'll have a nice press conference at the [bill] signing and that will be it." The PUC was required to establish a reasonable wholesale price for the duration of ten years. The price was expected to be in the range of 3 – 6¢ per kwh, important because the final price that was set could possibly decide whether the program would run out of money before the target was reached. If the production price was in excess of the set price level, it would be eligible for extra compensation from special funds that had already been collected from the consumers. The special funds were scheduled to be used for conservation, renewable power, energy research, and other public purpose programs. If the contract price set by the PUC was too low, the unavoidable result could be the early exhaustion of funds. The portfolio goals were not mandatory; therefore, no one could estimate what the price for the first round of contracts would be until about mid-2003. Experts had been expected to go before the PUC to argue for the lowest possible prices, while the generators would be before the PUC arguing for the highest possible price.

On September 23, FERC's administrative judge Curtis Wagner ruled that El Paso Corporation had deliberately squeezed customers off its pipeline system in order to drive California's prices up. If the commission accepted Wagner's decision, the penalties could range from several million dollars to more than $2 billion. El Paso labeled the decision "inappropriate and without precedent" and entreated the full commission to reject the decision. Davis's spokesman said, "FERC has all along been hesitant to go to lengths they did today. In that sense, it's a radical departure, but . . . unless they do something about it, it's just rhetoric." Michael Shames, a consumer advocate, opined the commission was known to respect Judge Wagner and it was "highly likely" that FERC would adopt his recommendations. The bigger issue, according to Shames and the utilities' attorneys, was how harshly the commission would act against El Paso in the next twelve to eighteen months. Shames said, "Is this going to be a painful experience for El Paso, or will it continue to get light-on-the-wrist slaps that have been the hallmark of the FERC?" The El Paso case was extremely critical to California as the state used vast amounts of natural gas. Wagner observed that El Paso was operating through two of its subsidiaries, El Paso Natural Gas Company, which operated the pipeline system supplying California, and the power-trading firm, El Paso Merchant, which held the rights to use considerable pipeline capacity. A Washington- based attorney for SCE said, "It's the first time that any federal official has made a decision that the conduct of any party actually caused the price increases in California. It's a very, very significant victory." According to PG&E attorneys, El Paso Merchant Energy made profits that ran into the hundreds of millions of dollars as prices rose. California's regulators had been attempting for two years to convince FERC that the two companies had a cozy relationship. Wagner ruled that El Paso's pipeline company had withheld "extremely large amounts" of pipeline space by running its system at about 79 percent of its design capacity. The practice violated both federal rules on how pipelines should be operated, including the previous settlement with El Paso in which they had promised not to cut supplies to California. In addition, El Paso's actions had tightened supplies and driven up California's wholesale prices to a point were they were much higher than in surrounding states. The judge urged FERC to hold penalty hearings on El Paso for violating FERC standards of conduct and unlawfully exercising market power. El Paso contended the judge misunderstood its system and its operating criteria and was now trying to "second-guess" the decisions that had been made for operational and safety reasons. El Paso blamed the higher prices on high demand and the shortage of natural gas. It further stated, "The evidence in this case demonstrates that, at all times, El Paso Natural Gas operated its systems to maximize the amount of capacity available to California." However, Wagner's ruling took many of El Paso's contentions apart piece by piece, detailing instances of "nonessential" maintenance and failure to allow all of the buyers and sellers to know how much space was available on the pipeline.

On September 27, state Treasurer Angelides presented a timetable for the sale of $12 billion in energy bonds. The sale would repay the State's general fund $6.2 billion and provide $3.5 billion to be used to repay bank loans and reimburse the cost of the bond sale. The bonds, originally scheduled to be on sale in July 2001, were delayed due to opposition from legislature Republicans. Another cause for delay was the PUC's rejection of a plan designed to reassure dedication of rate income for repayment to the bondholders. The bonds were to sell in two batches, variable rate bonds the week of October 28 and fixed rate bonds on November 1. The State was expected to place in the marketplace $25 billion of bond sales and revenue application notes in two to three months. The rating agencies insisted on adequate reserve to assure protection against unknowns and possible continued legal challenges to the utilities.

The passage of AB 6x prevented the utilities from selling any more of their power plants before 2006 and further provided that any facility sales would require the approval of the PUC. PG&E had not opposed AB 6x at the time but now argued that the 1996 deregulation had been violated with the passage of AB 6x. PG&E filed a suit in the Sacramento superior court contending that the 1996 deregulation law guaranteed the utilities the right to sell their hydroelectric plants or shift them to an affiliate. They also sought the right to sell their own generated power on the wholesale market under FERC rules that were more generous than state laws. A PG&E spokesman argued, "We lived up to our end of the bargain and the State ultimately didn't live up to its end but it attempted to change the rules after the fact." The State disagreed, and in March the Victims Compensation and Government Claims board formally rejected PG&E's breach-of-contract suit. The energy generated by PG&E's hydroelectric plants was some of the cheapest energy in California and could have been especially profitable. PG&E was not allowed to enjoy those profits as the PUC restricted the hydro retail price to charging the consumer the cost of operations plus a reasonable PUC designated profit. PG&E had wanted to be able to sell that hydro energy on the open market overseen by FERC, which allowed the charging of higher prices as long as they didn't abuse market rules.

October

State Treasurer Angelides finally realized the sale of his energy bonds. The sale of $4.25 billion bonds occurred on October 23, and it was anticipated that another $7.7 billion would be sold in two weeks. The twenty-year bonds contained an interest rate of 1.8 percent, slightly below the then current market rate. Credit agencies rated the bonds as "investment grade" and required sizable reserves to protect against uncertainties like electrical price spikes and litigation.

The PUC, on October 24, decided the means by which the different utility ratepayers would pay for the energy bonds and at the same time denied an exemption for those ratepayers who only used the 130 percent baseline power minimum. A bond charge was expected to be initially between .5¢ and .75¢ and would remain on utility bills for twenty years. The final amount was to be determined later in the year and could vary in the subsequent years. A formula for calculating it was established by a PUC vote. The PUC agreed to exempt special low-income and medical rate residential customers. The legislature already exempted those households from most of the 2001 rate increases. While there were some who believed that those residential consumers should have been included, the majority of the commissioners believed it would be fairer to exclude them. The commissioners stated that rate increases imposed in 2001 would cover all costs for PG&E and SCE customers, and those electrical bills would temporarily remain unchanged. The bond repayment decision was a necessary step in order to permit PG&E and SCE to take over the purchasing of power from the DWR. Included in the PUC decision were new rules determining how the utilities would buy power; additionally it required them to do long-range planning regarding how they would meet the customers' needs. The PUC postponed until November 2nd key issues, including how big a share of costs should be borne by those customers seeking direct access. As of January 1, 2003, the three utilities would assume the responsibility under DWR's long-term contracts to supply power to their customers. In PUC approval for future contracts, the utilities would be given some latitude in their general strategies, but contracts were required to meet their long-term strategies. The PUC would hold the utilities accountable if they failed to balance the use of their own generating facilities and the long-term contracts to create the lowest overall price for the consumer. PG&E spokesman John Nelson said, "They're trying to wave a magic wand and put the utilities back in the business of buying power, without addressing the fundamental concern of the utilities' financial health and investment grade status."

FERC, on October 30, said that it would meet on December 2 to hear arguments over whether or not El Paso Pipeline had schemed to drive up California natural gas prices by keeping massive amounts of gas off the market. FERC had the authority to fine El Paso and order refunds to the extent that El Paso profited by that alleged manipulation. Due to the nature of the judge's findings and the possible huge economic effects, as well as the possibility of the case ending up in federal courts, FERC decided to hear from the companies themselves before ruling. Pat Wood, FERC chairman, said, "This was an important case; we want to make sure we get it right." He added, "Hearing these en banc [by the full court] was not unprecedented, but it's a step that is seldom taken."

November

City-operated municipal utilities were enthusiastic supporters of renewable energy (green power) as long it was others that used that type of energy. Nancy Radar, a Bay Area energy consultant, said, "The fact is they [the munies] are not pulling their weight when it comes to promoting renewable [energy], and they benefit as much as anyone else from clean air." Green-power sources contended that if the current trend continued, power production nationwide could create more pollution by being vulnerable to price swings. That risk existed because for a long time the municipal utilities faced less state regulation than the three-investor utilities. The munies jealously guarded their status and fiercely defended their independence. Government-operated and cooperative utilities had been successful in lobbying to be excluded from a 10 percent renewable-power standard, which was being considered as part of a federal energy bill. California's municipal utilities, along with irrigation and rural cooperatives, were able to win exemptions from the 20 percent renewable-power quota. In the continental United States, the exemption trend was pervasive, as municipal utilities aggressively lobbied for exclusion from their state's mandated quotas. Julia Levin, California policy coordinator for the Union of Concerned Scientists, said, "It frustrates me when environmental groups and the Green Party just sort of blindly support public power and assume it's going to be cleaner; it isn't necessarily." California's municipal utilities were initially included in an early draft of the state's renewable energy bill, and they went ballistic. Ultimately their requirement was reduced to allowing them to set their own renewable energy standard, even if that standard was zero. The municipal utilities and cooperatives argued they deserved special treatment because the customers owned them. APPA vice president Joe Nipper said, "The primary reason why municipal utilities in particular have done relatively well throughout the western energy crisis has been [their] local control." The public utilities, in contrast, were regulated by the State who establishes both the rules for their behavior and as well as setting their rates. If a new program was imposed on them, the utilities could pass on the increased costs in their rates and generally with a profit. Despite that, according to the munies and green-power advocates, the munies had made significant strides in promoting energy efficiency along with clean, renewable energy.

November 7, the PUC decided who was to receive the lower rate levels for electrical energy. In a three to two split vote, lower rates were assigned to large businesses and institutions. The PUC included in the decision a surcharge designed to preserve the affordability of "direct access." However, preceding the vote there had been a contentious debate, as some believed the surcharge rate represented a loan to direct access customers. Beginning January 1, direct access customers would pay a special surcharge of 2.7¢ per kwh. Commissioner Jeff Brown predicted the 2.7¢ rate would increase the following year. Commissioner Brown believed the PUC had few choices if it wanted the customers to have the ability to choose their electrical suppliers, and the ability to choose a supplier had been a key feature of the deregulation. Brown said, "If actual costs are imposed, direst access would become uneconomic." Killing the program now would send "a terrible signal to the business community." Since no new customers had been able to sign up in the past year, the established rate would be paid by the existing direct-access customers. A majority of the biggest power users included retail chains, manufacturers, and several schools. These customers made up an estimated 10 percent to 15 percent of the state's electrical demand. The small-consumer advocates decried the surcharge as being far to low, while the big-power users praised it. Lobbyist D. J. Smith, whose firm represented the California Large Energy Consumers Association, said "The 2.7¢ appears to be fair . . . while not providing such a shock to energy-intensive commercial and industrial customers." The surcharge would not apply to those customers that had never switched back to the utilities but had remained with their direct-access suppliers; therefore, there was a minority of new direct-access customers. The surcharge was expected to raise about $500 million annually, but estimates indicated that direct-access customers were already more than $1 billion in arrears in paying their surcharge bills. No one knew what the surcharge amount should be to immediately catch up those direct-access customers. Various commissioners' estimates ranged all the way from 4¢ to 10¢ per kwh. Lobbyist Smith said, "It's a delayed payment, yes. It's like a mortgage. But it's not a subsidy. We will pay the entire costs, eventually." PUC president Lynch maintained that it was equally unrealistic to call the arrangement a loan because loans are made between "two willing parties," not one group being forced to pick up another's tab.

Mike Florio of TURN supported Lynch, saying it would be easy for a future commission to flip-flop and lower the surcharge before the large users had paid their share. Illustrating that risk of flip-flop, the PUC in another action reversed a pledge it had made in the winter/spring of 2001, when it raised the rates approximately 40 percent and said the money would be devoted only to current and future power costs.

On November 8, PG&E filed its three-year budget plan with the PUC. It requested a 4 percent increase in its natural gas rates and asked that any potential electrical rate cut be delayed. The request would result in annual income increase of $105 million for gas and $508 million for electrical energy. PG&E said, and consumer groups acknowledged, that the electrical-rate request would not trigger any rate increase. The consumer groups declared that was due to the 40 percent increase the utilities received last year. Mike Florio, representing TURN, said that if PG&E emerged from bankruptcy as a result of the PUC plan, rates would be brought back in line with costs, but the PG&E request would result in boosting rates back up again. PG&E said the purpose of the additional revenue was to keep electrical and gas systems safe and reliable.

On November 18, Michele Markey, a former official for the gas industry journal, *Platts Gas Daily*, testified before state senator Joe Dunn's (D-Garden Grove) committee that was investigating the energy market. Her testimony was with the understanding that she had been granted legal immunity. She testified that natural gas traders for large energy companies frequently misreported information to *Gas Daily* to spike the benchmark in their favor. She stated that during the energy crisis, generators and utilities blamed high electrical rates on inflated natural gas prices that had been falsely skewed. A file of documents that had been ammunition for misdemeanor charges against power generators, if they had knowingly rigged the index, supported those statements. Dunn said, "It seems to us that we have another example of pure, unadulterated manipulation of prices to the detriment of California consumers. If we accept the testimony today at face value, I can say without reservation that there was criminal conduct in the pricing of natural gas." Markey testified that it was common knowledge that companies exaggerated the volumes and/or prices of gas they sold in order to skew the prices in their favor. Then energy traders, in actions similar to the commodities market, could take "long" or "short" positions on the energy market. Traders were then enabled to increase profits by forcing the index either one way or another.

She testified that Enron, over a three-day period, managed to create an artificially low price index, and that omission was overlooked by the *Gas Daily*, which resulted in a situation that potentially created a $100,000 increase in daily profit. In another incident, two companies reported the same volume and price within fifteen minutes of each other, suggesting a "wash" transaction—the two companies traded natural gas back and forth for the purpose of manipulating the price index. She said that as a matter of policy, the *Gas Daily* deleted email messages and other records after two weeks. By 2001, the publication had received numerous complaints that its indices were inaccurate because of tampering. Based on those concerns, *Gas Daily* considered hiring an auditing firm to investigate its index-keeping practices, but when Platts took over the publication, the auditing plans were dropped. Markey claimed Platts faced pressure from Enron, which spent heavily on Platt's services, to drop the audit.

December

A FERC judge, on December 19, ruled the disputed long-term contracts should be upheld. FERC's administrative law judge Carmen A. Cintron said the utilities had not established that the contracts were either "unjust or unreasonable." This was despite another FERC judge's preliminary ruling in the previous week, which said California had been overcharged $1.8 billion by the energy companies during the energy crisis. The final decision would eventually be made by the FERC. California, Nevada, and Washington utilities maintained that the contracts were "unjust and unreasonable" because they had been negotiated under duress during peak spot market energy prices. California and its utilities, in a separate and larger case under consideration by FERC, argued there was a direct relationship between the long-term energy prices and California's electricity market. Erik Saltmarsh, acting head of CEOB, said that California never expected a Nevada decision to forecast what FERC might do on California's claims. He stated, "The two cases that were put on were very different." Power sellers believed that the judge's ruling indicated FERC was leaning toward California's claims, given the high standards of proof needed to set aside certain energy contracts. FERC had taken into consideration complaints by the PUC and CEOB that sought cancellation or reduction of about half of the $43 billion contracts the DWR had signed.

In a series of decisions released on December 19, a "critical milestone" was met by the PUC in the California energy crisis. The state's utilities had been ordered to resume the power buying business, a function the DWR had performed for the past two years. The DWR's purchasing authority terminated December 31, 2002. The utilities also had been ordered to resume both their former electrical energy usage and planning decisions. The PUC also issued guidelines on how the purchased power was to be coordinated with its own generation so as to give the consumer the lowest energy cost. In a contentious three to two decision, the commission established how much the utilities should be penalized for disregarding this directive. Approval had been still required from the bankruptcy court, and the commission was confident the court would give its blessing.

With all the legal elements properly in place, the DWR initiated the process of turning over major portions of its energy-buying function to the state's three major utilities. The DWR would retain legal ownership of the contracts, including meeting the obligations that related to the buying and selling of electricity—including making sure the energy suppliers performed in accordance with their contracts and including the collection of money from the utilities for payment to both the suppliers and the energy bondholders. The utilities, in turn, would resume their obligations of physical handling of the power itself, including deciding which contracts would be tapped for power and how that power would be coordinated with their own generation to give the consumer the lowest energy cost. PG&E spokesman John Nelson said, "That's our traditional function; we are willing to take that over."

The DWR had been originally selected for the function of buying power for the utilities because it already had a trading function. To accommodate this new ongoing function, they established a group designated California Energy Resources Scheduling, CERS. This was the office that actually negotiated the $43 billion of long-term contracts, of which twenty had been renegotiated. The CERS office employed a staff of about 130 and under these revised operations, it was expected the staff would be reduced to seventy to eighty employees. The CERS office was still renegotiating some long-term contracts and was to submit its financial needs to the PUC. Bill Marcus of JBS Energy observed this new division of responsibilities was expected to cause some efficiency with the utilities now coordinating their own power generation with purchased power. But he also noted there would be the possibility of increased costs, as the CERS staff requirement seemed excessive for the work that remained: "We're going to end up with duplication between the utilities and DWR." The DWR told the PUC it would need $5 billion in 2003 to repay bonds, fund the CERS' staff, and pay for energy bought on long-term contracts. Of that $5 billion, $28 million would be needed to fund the CERS staff, purchase supplies, and rent office space. All of the CERS expenses would be paid through the consumers' electric bills.

VI. The Witching Hours—Year 2003

January

FERC administrative law judge Bruce L. Birchman ruled that the power companies overcharged California during the 2000-2001 energy crisis. Subsequently he issued a judgment of $1.8 billion in the State's favor, which was considerably less than the $8.9 billion the State had been seeking. The State's initial unpaid balance to the power suppliers was $3 billion and the State ended up with a $1.2 billion unpaid balance. The judgment still required a review by the three-member commission. Concurrently FERC was in the process of reviewing charges that California's electrical and natural gas prices had been artificially jacked up. A favorable court ruling on natural gas price manipulation was crucial to the production of electricity and could return $1 billion to the State, according to Governor Davis.

February

Reliant Resources, after negotiations with FERC, agreed to pay California $13.8 million in damages for a June 2000 episode in which it deliberately withheld power in order to manipulate electrical prices. While the ruling was the largest agreement that FERC had approved to date, California dismissed it as ridiculously low; Governor Davis deemed it "insulting." Senator Bowen said the details of the two-day manipulation episode made "an absolute farce" of the generators' long-standing claims that they had not held back power to raise prices. Reliant called the incident an isolated occurrence that did not reflect its operating standards. They pointed out that they had alerted FERC to the incident themselves after conducting a review of records as ordered by the commission. FERC commissioner Nora Mead Brownell said FERC was still investigating the western markets and believed it was too soon to conclude how widespread those tactics were and what penalties the generators would be required to pay. According to Commissioner Brownell, "We need to keep in focus. It's two days, one company, and a clear, clear record of inappropriate behavior." She went on to say Reliant's behavior was so blatant that details of the violations should be released publicly. FERC did just that and released Reliant conversations that included the following snippets: "We decided prices were too low on the daily market, so we shut down everything except Ormond . . . Excellent. Excellent. That's sweet . . . Isn't it fun when you can do things like that now? . . . We literally shut down everything off but Ormond. Everybody's like, 'Watch us.' And it worked . . . Did the market find out? . . . No, God, no . . . That was fu-un."

The shutdowns were normally required to be reported to the ISO and not publicly disclosed under market rules in place at the time. The investigation revealed that the manipulation incidents occurred June 21 and June 22 of 2000. The timing was just as the crisis was getting into full swing. Prices had begun drifting upward a week before the first rolling blackouts that resulted from high temperatures and local supply problems. "Roughly 8,000 to 10,000 Mw statewide disappeared from the PX market, virtually overnight," said Mike Florio, a consumer attorney sitting on the ISO board. During 2000 and into 2001 an unusually high number of maintenance problems had appeared, and Florio cited those as the sham reasons for shutdowns. The industry blamed the breakdowns on the age of the plants that had been overworked in the early part of the crisis. Had those incidents not occurred on June 21 and 22, FERC concluded, wholesale energy buyers, including utilities, could have paid $13.8 million less for power. The commission that ordered Reliant to refund $13.8 million to the purchasers also required Reliant to hire an independent consultant to monitor all shutdowns during the next two years and to report on shutdowns to the commission. Additionally, the order included clauses that could force Reliant to sell power to California under specific conditions. Erik Saltmarsh, general counsel and acting head of CEOB, opined that Reliant apparently wanted high prices in late June in order to boost its second-quarter earnings by stating a higher value for contracts for the year. Such strategies were typical of high-level maneuvering in the state's complex power markets, wherein traders would lose money in the short run but benefit greatly over time. He went on to say that the settlement held shreds of hope for California because it might bolster the State's case for seeking refunds back to the year 2000 and could indicate that FERC might set more generous formulas for calculating the damage done by market tricks. "If FERC comes out with about five hundred more of these, then we'll be getting somewhere," Saltmarsh concluded.

March

Monday, March 3, a joint filing presented to FERC by the California attorney general's office, the PUC, the CEOB, PG&E, and SCE, demonstrated that the manipulation and price gouging of California's energy markets had cost the State $7.5 billion. Governor Davis told reporters the filings established "an industry-wide pattern of stealing and cheating California. It's just not a smoking gun; it's an entire arsenal." The California group believed that there was sufficient material to compel the FERC to rethink its decision that the State was entitled to only $1.8 billion. The report said, "The California parties have unearthed far more than enough factual information to warrant commission action." The report went on, "Even where the commission finds individual facts or numbers open to interpretation, it is inconceivable that it can now ignore a pattern unprecedented, unjust, and unreasonable prices and a pervasive, widespread and amply documented pattern of behavior by many sellers." The filing did not imply the State was backing away from its demand for refunds totaling $8.9 billion. The group said the manipulation they uncovered was most certainly more extensive and that if they had more time and better cooperation from the generators from whom they requested information, the refund figure likely would have been larger. FERC's spokesman Brian Lee said the State's information would be considered along with other data available. The commission was expected to commence making decisions regarding refunds that month. FERC's history of performance indicated any decisions would involve more hearings and could take several months before there was a final determination. Any of the commission's decisions were anticipated to be appealed in the federal courts. The Court of Appeals for the Ninth Circuit had sided with the State last fall in its claim that the commission had improperly limited the scope of its investigation along with its refusal to consider broader evidence of manipulation. The commission gave the State an additional one hundred days to develop more evidence of market manipulation. Erik Saltmarsh, along with others, said the one hundred-day extension was too short to be exhaustive and in many instances, the power generators either refused to submit the requested material or submitted such massive quantities of material that only a small portion of the material could be examined. The generators cited included Williams, Dynergy, Mirant, Reliant, Duke, Powerex, Coral, Sempra, and several other small companies. Saltmarsh said, "We can show pervasively that a number of major sellers were engaging in withholding power from the market." The report noted many of these companies had subscribed to a commercial service offered by Industrial Information Resources [IIR] that provides nonpublic information regarding who was withholding power and how much. It said, "The sellers receiving and sharing competitor outage information from IIR [Industrial Information Resources] then used this information in setting prices and making decisions as to where and how much to sell." The power generators denied the State's accusations, stating that many factors influenced the electricity crisis. In a press release, Duke said it planned to "formally and substantially refute any allegations against the company's California operations when it files its response to the California parties' allegations on March 20. Duke Energy looks forward to the opportunity to finally and fully defend the company's performance in a process based on facts, not rhetoric."

FERC and El Paso Corporation reached an agreement on March 20 for El Paso to pay $1.7 billion to settle charges that it had manipulated natural gas prices during the energy crisis. Included in the settlement were $900 million worth of natural gas over a twenty-year period and cash to utility ratepayers, various Southern California cities, the states of Nevada, Oregon, and Washington, and some private entities. California had been seeking $3.7 billion dollars for El Paso's market manipulation but hailed the agreement as proof that the energy crisis was, indeed, the result of market manipulation. Governor Davis said, "This is more evidence that the energy crisis was not a California-created problem, but rather market manipulation by the energy pirates that FERC was ready to sweep under the rug." Attorney General Lockyer added, "El Paso played a key role in the multibillion dollar rip-off of California committed by energy companies." The agreement was an agreement-in-principle and wouldn't be finalized for some time. UCEI's Borenstein said, "This is not just a token."

Paul Carpenter, energy consultant for SCE, said, "I think it's a fair number." The agreement stemmed from a three-year old FERC investigation into El Paso's alleged role in driving prices up during the energy crisis. California's PUC had filed a complaint with the FERC that said El Paso's mischief had cost the State $3.7 billion in direct higher costs and $2.7 for higher electrical costs stemming from the higher natural gas costs. The PUC argued that El Paso's auction had been rigged to favor El Paso Merchant Energy, causing the natural gas prices to jump from $3 per million BTU to $59 per million BTU in the ten months after the awarding of the pipeline contract. El Paso argued it had done nothing wrong, blaming the spike in gas prices on a gas pipeline explosion in New Mexico and subsequent government restrictions on pipeline deliveries. In addition to supplying free natural gas, El Paso was required to pay California and other entities $665 million in cash and the balance to be paid over twenty years. Of that, $505.6 million was dedicated to the ratepayers to be distributed under a PUC formula not yet determined, and $159.4 million was to be split among Nevada, Oregon, and Washington, two oil companies, and six Southern California cities and counties. El Paso would pay the State from an executive pool; $2 million to California, along with a decrease in its long-term contracts of $125 million.

El Paso Energy reached agreement with the CFTC to pay $20 million for its false reporting to the natural gas publication. The company agreed to make an initial payment of $10 million with the balance to be paid over the next three years. However, the company wouldn't admit to any wrongdoing. CFTC commissioner Sharon Brown-Hruska said, "The gas and power markets are vital to the U.S. economy, and the commission, as a market regulator, is committed to maintaining their credibility. This settlement demonstrates that commitment by sending a strong message that fraudulent and manipulative schemes will not be tolerated." El Paso on March 26 reaffirmed that it was cooperating with investigators and articulated that "any inaccurate reporting is unacceptable." El Paso Corporation announced it would disband El Paso Energy as an economy move. Internal reviews initiated the preceding year produced examples of other companies participating in fake reporting, including El Paso Corporation, Dynergy, Williams, CMS Energy, and American Electric Power Company. Dynergy was the first of those companies to reach a settlement with the CFTC in December, its total being $5 million.

FERC seemed to be moving in the right direction to resolve California's demanded damage awards, including bringing the energy marketers to task for their market mischief. FERC demonstrated action on March 26 by awarding refunds up to $3.3 billion. FERC singled out twelve companies for discipline and issued an exhaustive explanation of market exploitation that had occurred, leaving open the possibility of additional probes. The FERC did indicate that it was unlikely it would set aside or reduce California's long-term contracts, an issue that would be taken up later in the spring. Governor Davis said the commission lacked "true grit" to refund the full $9 billion the State had been seeking, but acknowledged its rulings did give the State a better chance in the federal courts. FERC chairman Wood said, "This commission is acting to ensure that customers pay just and reasonable prices. Today's actions represent important progress toward a just resolution of these matters for both customers and industry. It is time to bring this crisis to a close." One big victory that California welcomed was FERC's decision to use a different refund formula that would take into account the manipulation of natural gas prices in calculating the State's refund. The new formula, according to FERC officials, could boost the previously recommended $1.8 billion award to $3.3 billion. The final figure would not be known for months and could take until late summer to resolve. Wood opined the refund could possibly creep higher by millions, but not by billions. The State and others believed once the FERC reviewed the issue, they would eventually arrive at a much higher figure. The report also maintained that genuine power shortages had existed in California, and "a fatally flawed market design was the root cause of the California meltdown." Additionally, the problem was aggravated by the Pacific Northwest drought. The report named names and identified major generators, municipal utilities, and for-profit utilities as players that had extracted profits as a result of the state's energy crisis. Among other content, the report noted that (1) Enron made speculative profits of more than $500 million; (2) Reliant Energy Services sometime traded gas as frequently as every ten seconds, driving the prices up and down to reduce its own gas costs in a way that savaged the broader market, which added as much a $1.1 billion to Californians' natural gas bills and $1.6 billion to their electric bills; (3) Reliant and BP Energy coordinated their efforts to increase prices in 2000, with trades that commissioners found so worrisome that on March 26 they gave the companies three weeks to defend themselves or lose their ability to sell at market rates; (4) dozens of companies, large and small, engaged in market games similar to those colorfully named in the now-notorious Enron memos; included among them was the LADWP; and (5) PG&E and SCE had helped aggravate the market chaos when they tried to drive down prices, and PG&E sometimes lost control of power from its own plants to traders who then resold it at huge markups.

At the March 26 meeting, Mr. Wood concurred with commissioner Nora Mead Brownell in voicing opposition to abandoning the long-term contracts. Her position was that she believed in the "sanctity" of contracts and abandoning them would weaken confidence in the energy market. "This is a most difficult issue; the record is deep and mixed. The totality of circumstances on long-term contracts would suggest far more damage would be done to abrogate them." The lone dissenter was commissioner William Massey, who observed, "To me, there is absolutely no question that manipulation described by staff as epidemic had a huge impact on those long-term contracts. I'm inclined to believe that some of these contracts are going to have to be reformed in order to meet our responsibility of ensuring just and reasonable prices." California was successful in renegotiating about half of the $44 billion of long-term contracts, thereby reducing electricity costs by approximately $5.2 billion.

On March 27, FERC received a severe verbal dressing down from West Coast senators. Senator Marie Cantwell (D-WA) admonished, "The issue today is whether FERC is capable of doing its job." Senator Feinstein said she was "disgusted" by FERC's catalog of market abuses that constituted "one of the biggest frauds ever perpetrated on the West Coast." In a public discussion, the commission appeared to be split two to one against approving requests by the West Coast states for an order to rewrite the high priced long-term contracts. Senator Ron Wyden (D-OR) declared, "We've been hammered by overpriced contracts. Energy traders are on tape talking about manipulation. Somehow, for some reason, the commission can't see the connection between these smoking-gun acts and higher energy prices. We're talking about hundreds of millions of dollars. What more do you need for forward relief?" Commission chairman Pat Wood responded, "We've got further investigation to do."

In a March 27 hearing, the Senate Energy and Natural Resources Committee discussed whether Congress should approve FERC's single uniform market design for governing market rates coast to coast. Several members expressed skepticism about the design as a result of FERC's handling of California's electricity deregulation. Senator Larry Craig (D-ID) said, "My frustration is that you signed off on what was probably the most dysfunctional plan in the history of America, and now you are before us with another. When do you know what is right?"

Two weeks after announcing electrical refunds totaling $1.9 million, the FERC was set to order more refunds. These refunds were to come from companies that were willing to settle charges that they had manipulated the market.

To recap, in March FERC said California was entitled to about $3.3 billion for alleged overcharges during the 2000-2001 energy crisis, that is, the period between October 2000 and January 2001 when the State had started purchasing electrical energy for the utilities. California officials in turn challenged FERC and claimed $8.9 billion in overcharges; realistically, the officials doubted the State would ever get relief beyond the $3.3 billion. Apparently, the sellers of energy were willing to accept the $3.3 billion in refunds and indicated they would contest anything above that number. Although the $3.3 billion award was a partial victory for California, FERC was not sympathetic toward the other questionable time periods. In its investigation, FERC ignored the period between May 2000, when prices first soared, and October 2000. They stated that the sellers had not been put on notice until October of 2000, and that was when the agency could legally focus on specific acts of bad behavior. Saltmarsh said FERC ignored another $3 billion after January 17, 2001, when the DWR started buying energy for the utilities. FERC declined to examine DWR's spot-market purchases because, among other things, DWR's energy traders had been stationed on the floor of the ISO, which was a quasi-government agency. FERC maintained that the DWR's presence on the floor only gave it an insider's access to the wholesale market. California contended that the DWR had paid inflated prices equivalent to the prices paid by the utilities, and they would challenge FERC's position in federal court.

Davis's senior adviser Richard Katz said, "The governor has said he'll go to every court, every venue is the only process we have for getting justice for ratepayers." FERC decided that a considerable amount of electrical sales had not been eligible for refunds. Erik Saltmarsh, acting director of the CEOB, said "I am hugely disappointed that FERC's response is, 'We may give you 30¢ on the dollar, and you should be happy with that.'" Saltmarsh went on to say that FERC had estimated $4 billion in alleged overcharges and ruled them off limits for refunds; FERC took that position even though federal law required that wholesale prices be "just and reasonable."

In March William Keese, chairman of the CEC, drew the following sketch of California's energy situation during a speaking engagement with the El Dorado Irrigation District. He said California needed two 500 Mw power plants built annually to keep pace with the growing demand and that private companies didn't have the financial resources to build power plants. "No one is willing to loan them money, and no one except the munies and irrigation districts has the credit to borrow." The state could face a shortage of energy in the next three years, which would drive up prices. In the wake of the 2001 energy crisis, companies started building new power plants in the state; three were constructed in 2001, five in 2002, and seven more were scheduled to be completed in July. None were scheduled in the year 2003, and only one was scheduled for completion in 2005. Most of California's natural gas comes from out of state. Natural gas can be stored and the state generally enjoys an adequate supply, and more storage sites are being developed in California. Plans were underway for construction of liquefied natural gas, or LNG, terminals; five have been approved in Mexico, one for Los Angeles. There had been one planned for Mare Island in Vallejo, which was abandoned due to public safety concerns. He pointed out that liquefied natural gas is not the hazard the general public believes it to be; the gas will burn, but will not explode. The gas is liquefied to allow shipment in tankers. The state relies heavily on out-of-state hydroelectric power, particularly in the summer. He said about 10 percent of the state's energy comes from hydroelectric power. These plants are valuable in meeting peak demands as they can be put on line immediately. The CEC makes $140 million available annually in grants for renewable energy projects.

Attention at this time was also turned to the wind electrical generation industry, which depended heavily on the federal government's tax credits. A bill passed by Congress in 1994 credited wind generators with 1.8¢ per kwh lasting ten years. The opportunity to obtain the tax credit was two years and could be renewed by Congress. There were new wind farms in progress that were expected to be completed and producing energy before that credit opportunity expired. Seawest's senior vice president David Roberts said in many cases it took up to three years to plan and complete a new project and as a result, developers found themselves racing against time. "You have to take a risk that the tax credit is going to be there when you start the development of a project." The uncertainty of the availability of the tax credit significantly influenced the developer's ability to finance and secure a buyer for the generated power. "It's hard for wind projects to give a firm date of when they can be on line. The best we can promise right now is ten to twelve months after the tax credit is extended," said Scott Piscitello, development manger for Renewable Energy Systems. Developers anticipating the expiration of the tax credit opportunity in 2001 installed about 1,700 Mw of wind turbines, locking in a decade of tax credits. The following year the lawmakers in a miscue allowed the tax credit to lapse for three months, resulting in wind investment dropping to 410 Mw. Randall Swisher, executive director of the American Wind Energy Association, said, "By the time the credit finally did get extended, most of the projects were in limbo." Because of decreased demand for its turbines, Vestas American Wind Technology had in the last year shelved its plans to build a factory in Portland, Oregon. The company also cut six thousand jobs in the United States. Scott Kringen, marketing coordinator for the company, said, "What we're waiting for is a comprehensive energy bill that contains an extension of the production tax credit. We're looking for the market to stabilize and have a strong commitment to wind. It's going to be tough, but I believe it's not a question of if, it's a question of when." Federal lawmakers introduced legislation that would extend the wind tax credit anywhere from three to ten years. Kringen said Vestas was encouraged by the news of the legislation. He said a three to five year extension would be ideal for developers who need to get projects online without worrying that the incentive would evaporate. Time was running out for tax credit extension to have an effect in 2004. Dean Gosselin, vice president of FPL Energy, said, "We hope the production tax credit gets extended; without it, I believe the industry will be truly set back—if not completely eliminated." Swisher said a recent congressional review indicated the tax credits would cost $80 million over five years.

April

On April 9, the House Government Reform subcommittee, chaired by Doug Ose (R-CA), heard witness testimony on California's energy situation. The head of California's ISO reported that with the economy limping along and with the state's current facilities, there would be adequate power in the state into 2006 and 2007. But he warned the subcommittee all of that could change if the economy picked up or if the aging power plants were shut down or if the West Coast suffered another dry year. ISO chairman Winter told Ose, "If our economy turns around, it can add back [power demand] in a matter of months. If the economy picks up, we could be getting into trouble in 2005, 2004 with bad weather." He went on to say the basic market structure had not changed significantly and that it had dampened investors' interest in building any critically needed power plants and transmission lines. Karen Tomcala, PG&E's vice president of regulatory affairs, said, "The upheaval associated with the California crisis, not yet fully resolved, has chilled investment in the state." Jan Smutney-Jones warned that time was running out for a solution; he pointed out, "The energy crisis was a convergence of a serious supply-and-demand imbalance, poor market design, and an inadequate regulatory response. It need not be repeated." The ISO had proposed changes in presenting its plan, titled Market Design 2002, to state and federal regulators. It was designed to repair the flaws in the deregulation plan and prohibited the kind of market manipulation that had allowed marketers to drive up wholesale prices. Winters said, however, that it was now apparent that very little would be done until 2004 at the earliest, and even then it might not be enough. NCPA general manager George Fraser said, "I don't see it providing incentives for investment. There's nothing in it on transmission resources and adequacy." FERC Chairman Pat Wood observed that California relies on its neighbors for electricity and gas more than any other state does and warned that until California builds new generation and transmission lines and recasts its regulation of retail sales so that residential and business customers can respond promptly to rising wholesale rates through conservation, nothing can protect its customers from inevitable resulting problems. Winters said the ISO projected that the state would have 3,246 Mw of excess power available in the current summer, roughly the output of three nuclear power plants. "It looks like we're going to be able make it this summer. Beyond that, I have considerable concerns." Gary Ackerman of WPTF said there was additional generation available in Arizona and New Mexico, but it wouldn't be of much help without the transmission facilities to bring it into the state. Ose said all of this brought an unwelcome dilemma. "I want the economy back but if we get economic growth of 2.8 or 3 percent, what I hear you saying is we'll be in a box." Winter concurred.

State Senate bill SB 888 was unveiled on April 9. The bill would return California to regulated electric energy. Senator Joe Dunn (D-Garden Grove) commented that the details were still to be worked out and announced, "At no time will we compromise on our goal of ending energy deregulation in California. We aren't mending it; we're ending it." In essence the bill would reverse AB 1890 by including the permanent voiding of the "direct access" provision that had been in suspension, although there had been efforts to revive it. SB 888 would also extend the moratorium on the utilities' ability to sell more of their assets and would encourage investment in generation and transmission facilities. Dunn said, "What it does is return an obligation to serve to the utilities, in return for guaranteed cost recovery; it encourages utilities to return to generation." Senator Bowen said, "It has to happen this year because we need investments in capital facilities to be made beginning this year and next year, or we will have power shortages in 2006." Bill SB 888 supporters contended that by ending "direct access," the utilities could once again predict demand and would be able to enter into long-term contracts that might help keep wholesale prices down. Bowen went on, "We can have a mix of privately-owned generation, publicly-owned generation. It doesn't matter as long as those who are buying for a customer base can reasonably estimate." Critics argued that the bill would return to the old days of regulation and destroy the positive effects that had been realized from open markets. Jan Smutney-Jones of APPA said, "We basically believe that this re-monopolization will short-circuit California's energy future. This bill isn't really necessary. It's very troubling, and I guess we'll spend the summer arguing about it." The concern was that the bill would return the state to the old model wherein; supposedly, bloated utilities make wasteful decisions and stick customers with the bill. Severin Borenstein opined that getting rid of direct access would be a mistake and pointed out that California had some of the highest electrical rates in the nation even before deregulation. "To say we're just going back to the old system because deregulation sucks strikes me as not a very careful analysis of what happened." Assemblyman Keith Richman (R-Northridge) introduced a competing bill designed to preserve the direct access feature. It would require large energy users who wanted to enter into deals with producers to give adequate notice to the utilities. The bill further stipulated that the users were required to have a prior executed contract with a generator. If the user wanted an early return to its utility, the user would have to cover the

full cost to the utility that had been providing the power and probably at spot market rates. Richman said that if deals could not be made with a generator, the generator might choose to shut down its plants. PG&E warned against premature action, saying, "The last thing the state needs is to lurch off in a new direction without careful analysis and a full understanding of what the results will be."

May

On May 13, Governor Davis announced that the DWR would cut $1 billion from its budget for the year 2003 and indicated that "the rates are going down—this has been a very long ordeal." This cost reduction meant lower rates in the summer and in the fall for customers of SCE and SDG&E. PG&E customers would need to wait a bit longer as the PUC was hesitant to adjust its customers' rates because of the ongoing bankruptcy proceedings. The estimated reduction of the rates was 4 percent to 5 percent, with some customers getting nothing and others getting more, depending on how the PUC allocated the cut. These reductions would be in addition to SCE's proposed rate cut of 13 percent, scheduled to take effect in July. Consumer's advocate Mike Florio observed that the rate adjustments were headed in the right direction: "We hope it starts a trend." The DWR had asked for the extra large reserve in its budget for 2003 in order to reassure Wall Street that the DWR could step in to help if the utilities needed assistance after January 1 when the utilities started buying their own power. SCE and PG&E welcomed the news and had hoped that the DWR would quickly pass on the savings to the customers. PG&E spokesman John Nelson said, "We're a pass-through agent for the DWR. If they tell us 'We don't need as much from your customers,' then we can collect less from our customers to pay the DWR. That should be independent from any bankruptcy proceeding." Commissioner Susan Kennedy said, "There's very strong interest in getting the money back to the ratepayers as quickly as possible, but the PG&E bankruptcy created an additional hurdle." Commissioner Geoffrey Brown opined that while PG&E's lower rates went "on hold," he was confident that PG&E customers would eventually receive their share of the $1 billion rate reduction.

May 14, PG&E's reorganization plan, particularly PG&E's desire to circumvent PUC's authority, was being considered by the Court of Appeals for the Ninth Circuit. The court referenced a 1984 U.S. Bankruptcy Code amendment that permitted valid reorganization plans to take effect "notwithstanding any otherwise applicable non-bankruptcy law." The meaning of that statement was considered critical by some, as they contended that an expansive reading of the preemption phrase, as advocated by PG&E, could set aside regulations on everything from power availability to nuclear safety. PG&E claimed the fears were unfounded, arguing it required a broad interpretation to permit it to spin off its generation and transmission facilities. It would permit using those facilities as collateral in the borrowing of money without PUC approval. Two lower courts were split on the case. U.S. Justice Department lawyer Jeffrey Clair supported the State's interpretation of the phrase, saying PG&E's version would remove the Nuclear Regulatory Agency's power to block the transfer of nuclear power plants to unqualified owners. Circuit judge Michael Daly Hawkins of Phoenix asked, "So what does 'notwithstanding' mean?" That prompted a response from circuit judge William Fletcher: "You would replace 'any' with 'some.'" Harvard professor Laurence Tribe, representing PG&E, argued "any" means "all" and went on to say some federal regulations might prevail anyway. Judge Fletcher rebutted, "When the competing law is federal rather than state, then 'any' doesn't mean 'any.'" The hearing ended without a resolution and no date set for resolution.

On May 27, TURN, represented by Michael Strumwasser, was in the California Supreme Court, challenging the PUC's authorization of $3.6 billion to cover SCE's debt. Oral arguments addressed two issues: (1) did the PUC "change" SCE's retail rates by canceling a legislated rate reduction, and (2) did the PUC violate the state law that required open meetings for the adoption of rate "changes"?

Strumwasser argued that the rate deal represented a "change," saying, "It freezes a rate that by law had to go down." He also argued that although the PUC was allowed to consult privately, the PUC should not be permitted to approve a rate increase privately. Justice Joyce Kennard asked whether maintaining a rate change would constitute changing it; she suggested the answer was no. Chief Justice Ronald George observed even though the PUC had voted in private, there had been open meetings before and afterwards. Justice Marvin Baxter was concerned that a broad exception to the open-meeting law could provide a road map for filing lawsuits in order to get secret settlements. He asked, "How far could the

PUC go secretly—for the next century?" A decision was expected within ninety days.

On May 28, the ISO declared the year's first Stage 1 power alert starting at 3 p.m. and lasting until 6 p.m.; however, no rolling blackouts threatened the state. The alert did raise some questions: were conservation efforts failing, or had an unrecognized growth compromised the state's power situation, or was it possible the ISO had misjudged the state's power requirements. The ISO's director of operations said, "We need to analyze this. I've got a whole slew of engineers going off to do some work tonight." What was troubling was the higher demand; the demand statewide peaked at 39,577 Mw about 4 p.m. versus a forecasted demand of 35,012 Mw.

June

On Thursday, June 12, the U.S. Senate voted its approval of a bill directing the Department of the Interior to study the extent of oil and natural gas reserves available off the West Coast, the Atlantic Coast and the Florida Gulf Coast. An amendment to block the exploration was initiated by Senators Feinstein and Bob Graham (D-FL) and was defeated by a vote of fifty-four to forty-five. They, among others, argued that the only reason for the exploration was to exploit those reserves. Senator Pete Domenici (R-NM), Chairman of the Senate Energy and Natural Resources Committee, said, "The provision does not lift the moratoriums; this simply directs the Secretary of the Interior Department to conduct a study. It is prudent to take an inventory of our resources and where they are located. Knowledge is better than no knowledge." Senator Mary Landrieu (D-LA) added, "This amendment doesn't have anything to do with drilling or lifting. It will simply produce information that will allow us to make informed decisions." Senator Feinstein replied, "The real intent is clear. California knows what's at stake, and we have made an informed decision. We do not want drilling off our coast."

Nationwide natural gas prices began surging upwards, increases which would affect those consumers that heated with natural gas. Approximately 85 percent of California's households relied on natural gas for heating. Electricity, on the other hand, was not expected to see big increases, despite the price increases for natural gas, due to the long-term contracts the DWR had negotiated and the state's available nuclear, hydroelectric, and geothermal power. The PUC and the DWR estimated that about 15 percent of the state's electricity could fluctuate based on gas prices. In the current month, natural gas utilities were charging more than double the prices charged the preceding June. Bill Marcus of JBS Energy said natural gas prices statewide and nationwide would be the next unpleasant surprise. The House Energy and Commerce Committee were told by the Federal Reserve chairman, "We are not apt to return to earlier periods of relative abundance and low prices anytime soon." R. W. Beck's gas price forecaster suggested that after six to eighteen months, natural gas prices would probably settle around $3 to $4 per MM BTU for the next decade. In the preceding year, natural gas prices ran around $2 to $3 per MM BTU and according to some analysts could reach $7.50 to $10 per MM BTU in the current winter.

The California senate barely passed SB 888, which included a final amendment that declared the bill "for display purposes only." Senator Joe Dunn (D-Garden Grove), author of the bill, said, "There is an urgency here. Every stakeholder, whether they believe in the deregulation of energy or not, agrees that if we don't do something now—this year—as early as 2004 we could be right back into shortages." There was no support at that time for deregulation within the senate. Some Democratic senators were concerned that the bill didn't contain adequate environmental protections, while others were skeptical of the bill achieving its desired goals. Many business interests believed that the legislative fixes presented to date should have been allowed to take effect. Dominic Dimare, a CCC lobbyist, said businesses were opposed to the bill, declaring that they wanted more choices and not a return to regulation: "The worst thing you can do is overcorrect. SB 888 is overcorrection." The bill would delay the customer's ability to shop for "direct access." Also, the bill included legislation that allowed the State to buy and sell power, permitted utilities to sign long-term contracts, and gave the State the ability to inspect downed power plants. UCEI's Borenstein said direct access was a causality of the bill: "It's unfortunate because direct access was not a problem; it was a symptom of the problem." He went on to argue that the bill fell short of returning the state to pre-deregulation status; instead it would create a new hybrid system in which utilities would still not produce all of their own power and customers wouldn't have the ability to shop around: "This bill is abandoning some areas where the market could be helping some customers." He felt the state still needed to address issues that were brought to light during rolling power blackouts and spiking prices. In his opinion, the bill was a "knee-jerk" reaction—"We need a more careful reaction to what went wrong."

On June 19, state regulators and PG&E reached a tentative agreement to remove PG&E from bankruptcy protection. The plan still required approval by the PUC and the federal bankruptcy court, and approval was expected by the end of the year. Under the plan, billions of already collected dollars would be used to pay off PG&E's debt of $12 billion along with a portion of future collections. The plan required PG&E rates to be reduced by .5¢ on January 1 followed by another .5¢ in 2008. By contrast, SCE expected to reduce its rates by 2¢ in the coming summer, to be followed by another .5¢. This had resulted from DWR's request for less money in its midyear adjustment. PG&E agreed to give up its efforts to create new, less regulated companies for the next nine years and promised it wouldn't fight state law that required PUC review of any future sales of its assets. PG&E contracted to dedicate 140,000 acres of watershed land around its hydroelectric power plants (an estimated value of $300 million) for public purposes. PUC president Michael Peevey pledged to retain "an open mind" during deliberations, adding "the most important thing about all of this is that it has kept PG&E intact as an entity under state regulation." Consumer groups immediately condemned the plan, and some commissioners, along with Governor Davis, were highly skeptical. TURN's attorney Mike Florio said, "Either it's got to be a complete kangaroo court phony process or there's got to be a possibility of making the deal better. My suspicion is there are two deals. There's the real deal and then there's the fake deal they can put out there and get shot full of holes."

FERC on June 25 ordered administrative trials for about sixty power-selling companies. FERC indicated the maximum penalties these companies could face were the return of profits for gaming the market. FERC did reject California's demand for renegotiation of about $12 billion in long-term contracts. The rejection only affected about a third of California's contracts as the State had already renegotiated twenty-four contracts. FERC also rejected refunds for the Pacific Northwest that had resulted from the energy crisis and ended California's claim for $1.5 billion from the BPA, Powerex, and others. California deputy attorney general said, "The commission has completely ignored evidence that manipulation in California greatly impacted the Pacific Northwest. To get nothing is incredulous." Commissioner Massey had several concerns contrary to the majority of the commissioners. He noted that the commission was limiting punishment before it knew the extent of the misconduct. He had the same complaint against the commission's decision to review hundreds of thousands of bids submitted by suppliers in 2000 that were above the target price of $250 per mwh. He also noted that several power sellers were publicly-owned utilities that were supposed to be exempt from FERC's authority. Wood thought the publicly-owned utilities could be included because they were buying and selling power through the state's ISO over which FERC did have authority.

July

Shell and Bechtel were going to collaborate on an installation of a liquefied natural gas (LNG) facility at Mare Island in Vallejo, California, but the residents had serious concerns about having a potential hazardous facility in their neighborhood. The U.S. had developed an insatiable appetite for natural gas, and its ability to produce enough of this resource within the U.S. appeared to ending. Federal Reserve chairman Alan Greenspan told a congressional committee, "There is no way we can be self-sufficient." He reported natural gas was selling for double the price of a year ago, and supply and demand for gas was out of sync. In other words, he believed the U.S. would have to depend on sources outside the U.S., as it does now for oil. Sixty percent of the U.S. households depend on natural gas for heating, cooking, and hot water heating, and the demand is exacerbated by the nation's ever-increasing use of natural gas for electrical generation. The DOE predicted the use of natural gas would increase by 52 percent by 2025, while production of this resource would increase by only 35 percent. Robert Ineson, director at CERA, said, "We've decided we want a lot of electric generation from gas because it's cleaner, but we've thought it to be limitless and inexpensive, and that's proven not to be the case."

Production of LNG had increased overseas. Japan gets just about all its natural gas through LNG shipments. Apparently, approximately twenty LNG facilitates were currently under consideration as a result of a recently streamlined federal approval process, and the FERC had relaxed its anti-monopoly regulations for LNG. Cheniere Energy was planning four LNG facilities in the Gulf of Mexico as a result of acquiring two large customers, Dow Chemical and Sherwin Alumina. Cheniere's target was to be "the primary gateway for LNG in the United States." Their biggest hurdle, as Shell and Bechtel had learned, would be state and local approvals. The American Gas Association, representing the utilities, said even with imports, the current demand would require the opening of more federal lands for gas development.

On July 10, the Assembly Utilities and Commerce Committee rejected outright senate measure SB 888 that would return the state to a regulated market. Assemblyman Manny Diaz (D-San Jose) said, "I don't want to repeat what happened in '96 when a lot of legislatures rushed into something, and now we regret what happened. I just want to make sure we get it right." Power producers and many commercial electrical energy users had opposed the bill.

In the U.S. Senate, Senator Maria Cantwell (D-WA) offered an amendment to the Senator Pete Domenici's (R-NM) bill designed to outlaw energy market manipulation, but it was defeated fifty to forty-eight on July 29. Cantwell contended Domenici's bill, as written, left too many loopholes and did not require FERC to impose tough sanctions on those who broke the law in market manipulation. Republicans rebutted there was adequate protection in the bill, and California's energy crisis was not the result of insufficient generation and the state's faulty deregulation but of market manipulation. Cantwell went on to argue that the West Coast energy crisis had cost ratepayers in excess of $6 billion in their energy bills and had also cost the U.S. economy $35 billion, including the loss of 589,000 jobs. She continued saying that Congress ought to be careful in deregulating wholesale electric markets. California's two senators, Dianne Feinstein and Barbara Boxer, supported the amendment. Feinstein said if the law weren't tough enough, there could be a repeat of 2000-2001 episodes—"This could be California all over again, [and] no one is certain how this deregulation will affect energy markets." She went on, "These companies will try to do it again if they can. This should be dealt with. The fact of the matter is FERC has not dealt with it." Senator Boxer, in reference to California's budget crisis, added, "One-third of the debt our state faces is related to the energy scam." Domenici disagreed, saying, "What happened before won't happen again."

California targeted 2010 as the year to be generating 20 percent of its electrical energy demand from renewable energy sources. On July 29, state Power Authority chairman David Freeman threatened that if the PUC didn't require utilities to provide more electricity from renewable generation by the end of the year, he would seek authority from the legislature to direct the utilities to do so—"We're going to see some renewable action in this state, one way or another." He went on to say, "It's an important aspect of our energy plan. The utilities need to be entering into long-term utility contracts that will support the financing and building of new wind farms and solar power and other renewable energy." PUC spokeswoman Terrie Prosper responded, saying there would be no need for the authority to seek a legal change, as the PUC was already working to ensure the state would meet its renewable energy goal. California environmentalist Evan Paul commented that California's agricultural waste, wind-blown hills, and year-round sunshine make it an ideal place to develop industries in biomass, wind, hydroelectric, and solar energy—"It definitely rivals the computer industry for the amount of growth we're going to be seeing in the next ten to fifteen years."

August

The Utility Reform Network on August 21 lost its suit against the PUC when the California Supreme Court upheld the PUC's authority to bail out SCE and shift onto ratepayers $3 billion of costs for the 2000-2001 energy crisis. The ruling maintained existing rates. The court opined a reduction in rates hadn't been guaranteed; it emphasized that the PUC, by law, was responsible to see that the electric utilities had the capacity and financial viability to provide power to the state's consumers.

On August 27, the Bush administration signed a new law that allowed older power plants to be upgraded without pollution control additions, if the plant upgrades were less than 20 percent of the plant's value. The law increased tensions between those who wanted increased electrical energy production and those concerned about potential environmental degradation. A suit California had filed in February was to be joined by other states. Scott Segal, director of the Electric Reliability Coordinating Council, believed the change could enable the increase of coal-fired plants by 10 percent, which was equivalent to sixty average-sized power plants. William Becker, executive director of a Washington-based organization representing state and local air regulators said, "This rule guts one of the most important parts of the Clean Air Act."

September

On September 4, the PUC ordered a one-time discount on Californians' electricity bills, to be applied to customers of the three major utilities. A total of about $1 billion was scheduled for refund. The money was being refunded from a special contingency fund, which had been DWR's hedge in the event DWR's transition to the utilities did not go smoothly. However, the January 1, 2003, transition did go smoothly and the special transition fund charged to the consumers was found to be unnecessary. The refund was to be based on how much electricity each customer had used over the past twelve months. PG&E estimated that an average residential customer used about 500 kwh each month and would receive about a $40 refund, small businesses would receive about $125, agriculture customers about $300, larger businesses about $6,700, and the biggest industrial users would receive about $72,000. Michael Peevey, president of the PUC, said, "This is one more example, in our view, of putting the energy crisis behind us and putting money back into the hands of consumers." However, PG&E customers would continue paying the highest electrical rates in the nation under the proposed settlement of PG&E's bankruptcy protection case. The agreement between the PUC staff and PG&E had been under review by the PUC; it called for a .5¢ rate reduction in January with a similar reduction in 2008. The two cuts would eliminate 1¢ per kwh from the 4¢ increase imposed in 2001. By late September, Californians could expect a credit on their electricity bills.

On September 9, SMUD received approval from California regulators to construct its Consumnes Power Plant at the site of the former Rancho Seco Nuclear Generating Station, which had been decommissioned in 1989. Construction now awaited the approval of federal permits, which were expected to be available in October. The federal permit decision was to be based on whether or not SMUD had satisfied various regulations, including those relating to wetlands. The plant was scheduled to start generating in mid-August of 2005. The estimated cost was $400 million, an investment that SMUD had already factored into its existing rates. The project was designed for an initial 500 Mw. SMUD also planned to build a twenty-six-mile natural gas pipeline from one of its smaller plants in Elk Grove to the site. Representatives for several rural ranchers complained to the CEC that SMUD wanted to use more of their land for the right-of-way than they considered necessary. In addition, the plant came under fire from environmentalists who contended the project's wastewater could potentially pollute streams. SMUD changed the design of its cooling systems to mitigate cooling water discharged into the local creek, and they also pledged to consider the use of recycled water in the plant's planned second phase, which would eventually expand the plant capacity to 1,000 Mw.

Congress was considering energy legislation that would determine the fate of the seven-decade-old 1935 Public Utility Holding Company Act (PUCHA). The duty of enforcing the law had been assigned to the SEC, an agency that also considered the law obsolete. Both congressional houses had voted to ditch the law. The Bush administration advocated a provision in the energy bill to control holding companies, which, with no experience in the energy business, would be free to buy cash-rich utilities. David Owens, executive vice president of the Edison Electric Institute, told Congress in March, "PUCHA is a long-standing barrier to capital investment in the utility industry, creation of independent regional transmission companies, and the entry of additional players in the wholesale and retail energy markets." The purpose of the PUCHA, a Depression-period law, was to stop the fraudulent accounting practices of electric utilities that were creating huge financial losses for investors and placing severe rate increases on consumers in order to prop up failing operations of their non-utility subsidiaries. Under PUCHA, holding companies owning utilities were required by law to dedicate their business pursuit to the utility business. The holding companies were also barred from conducting business in areas not contiguous to their area of business, i.e., they could not own utilities in disconnected parts of the country. Accordingly, the act limited the holding companies' geographic reach of utilities across the nation. Industrial and financial leaders believed if the act was repealed, the industry could expect a rush to mergers, and the industry would diminish into a dozen or so large holding companies. The proposal did have senate opposition; Senator Pete Domenici (R-NM), supported by the APPA, believed that if the act was abolished, there should also be measures to increase the FERC's authority to review mergers. Mark Cooper of the Consumers Federation of America said, "If the PUCHA is dead, you watch. What happened in California will start showing up elsewhere." Critics believed that FERC did not have the experienced staff necessary to assume SEC's responsibilities and that the commission lacked the expertise and legal authority to conduct such reviews. NARC, however, supported repeal of the act as long as state commissions were given the authority to examine out-of-state holding companies' books.

The last major revision to the act had been made in 1992. At that time, the restrictions on holding companies had been eased in order to allow the holding companies to invest in building power plants without restrictions on locations and also to allow them to sell their power in the open wholesale markets at the current rates. The new freedom encouraged the construction of new plants, but it also initiated risky ventures and enabled companies like Enron to indulge in market gaming. Isaac Hunt, former chairman of the SEC, in a House hearing in June of 2001, testified that the 1992 PUCHA reforms had nothing to do with the California crisis. Doug Davidson, whose law firm specializes in PUCHA, said, "I see the prospect for even more dramatic changes in the industry than we've seen in the last ten years, but is this frightening? No." He said it was a good thing, as it would bring huge amounts of capital to build efficient transmission systems that the utilities had been reluctant to finance in the current climate. Davidson conceded there would be a regulatory gap with the act's repeal, but "there is some real prospect that the gap will be filled in part by expanding the jurisdiction of FERC and state regulatory agencies." Alan Richardson of the PPA questioned that prospect, citing the widespread financial failures of merchant energy companies after passage of the 1992 revisions to the act. "The one thing we can be certain is the law of unintended consequences will come into play."

Robert Burns, senior research specialist at NARC's research arm, the National Regulatory Research Institute, opined that with the climate of deregulation, the repeal of the act could be a tough sell, something that should be "approached with caution." He went on to say, "We're very concerned about market problems now. If we allow the industry to consolidate, it could make it more difficult to have any competition at the generation level, and we might find that we would be forced to over regulate all over again. It's not necessarily a bad idea. But if it's not done correctly, it can do more harm than good."

Calpine had been seeking to drill for geothermal steam in Telephone Flat, an eight-square-mile Medicine Lake Highlands area located 30 miles northeast of Mount Shasta. The area first attracted energy companies in the early 1980s, and the companies sought the drilling rights on the property. The Bureau of Land Management leased out the drilling rights, and the Forest Service, which managed the surface area, had no objections at the time. Objections, however, came from the Indian tribes throughout the western states. The Highlands were important to them, as they considered the area sacred for healing. Indians came to the Highlands to pray, gather medicinal herbs, and to learn the secrets of the Earth's creation. In July 28, 1998, the Forest Service declared the upper reaches of the mountain off limits to development. The keeper of the National Register of Historic Places designated the site as a national historic site. Calpine had acquired the drilling rights to the leasehold in the fall of 2001. However, there had been dissension over the development among the Shasta tribes, which split into two factions that remain to this day. The Pit River tribe took the view that jobs were more important than cultural values. In May of 2000, however, the preservationists gained momentum when the Bureau of Land Management and the Forest Service denied all drilling permits. The two government agencies placed a five-year moratorium on the remaining Fourmile Hill lease area until further studies of cultural and historic significance had been conducted. Howard Wyant of the Shasta tribe voiced, "The local Forest Service is our last line of defense."

That line of defense started to crumble when the tribes faced legal threats from Calpine and with the election of President Bush. Bush's administration reversed the ban on drilling and lifted the moratorium on Fourmile Hill, citing the need for development of domestic energy resources. In September the issue landed in Sacramento's federal court. Tribal members were joined by environmental groups and all called for a complete ban on geothermal development anywhere in the Highlands area. They cited the area for its cultural and spiritual importance and claimed that mercury, arsenic, and other toxic chemicals would be dredged up by the drilling operations and would contaminate a huge underground aquifer that fed nearby rivers. The opposition to Calpine's drilling plans was prepared to take the case to the Supreme Court, if necessary.

Calpine, in turn, maintained that drilling operations could be carried out in a way that would respect the Indian's sacred sites. Drilling operations would steer the actual drilling activity away from sacred sites. Additionally, the drilling operations would be muffled to minimize noise, and wells would be encased in concrete to mitigate the escape of toxic chemicals. Calpine suggested observers from various tribes could monitor the drilling operations. Calpine also argued that the area was not pristine, as it had undergone repeated timber cuttings in the past. The region was a popular tourist destination as well. Hundreds of vacation cabins dotted lakeshore and forest. Each year, the area drew an estimated 40,000 visitors. Water skiers enjoyed the lake in summer, and snowmobiles invaded in winter. Calpine, facing considerable local opposition, was obliged to wait the outcome of the lawsuits before it would be allowed to commence drilling operations.

Calpine estimated it had spent over $18 million in lease purchases, payments, and exploration, and it intended to recoup its investment, as it believed renewable energy was the wave of the future. The CEC had planned to invest a $20.8 million grant into Calpine's Fourmile Hill project if it became a reality. The grant was to be financed through a surcharge to the ratepayer. Calpine had originally been working on the basis of some encouraging test drilling results. However, it turned out that the findings were not sufficient to qualify the site as a commercially viable resource. CEC put $1 million into test drilling but concluded that the area's resources were not adequate to warrant development. Calpine reported operation losses since the previous fall.

October

On October 2, Reliant and FERC reached a $50 million settlement, the largest one, for mischief attributed to Reliant's five subsidiaries in their manipulation of the market. In addition, FERC's staff recommended a series of other settlements with penalties that amounted to less than $3 million. FERC maintained that the settlements were a result of FERC's investigative efforts and, therefore, were not subject to review by parties outside the investigations. FERC subsequently reported that it was dismissing all of California's objections to a series of settlements with electrical energy generators for their mischief in manipulating the California market over the preceding two years. William Massay, the lone Democrat on the commission, disagreed with the decision, contending that the commission's investigation had been based on evidence coming from California interests who had been denied a voice in the settlements.

The $50 million settlement with Reliant was questioned by California, which argued that the energy traders, in their illegal actions, had taken advantage of the state's power and natural gas shortages in 2000 and 2001 by dramatically increasing wholesale power prices. The elements of the schemes were similar to those revealed in an Enron memorandum, including the practice of shipping power out of state and then re-importing it, the withholding of power, and associated coordinated trading. It was argued these schemes exaggerated California's energy shortages and drove up wholesale power prices. FERC's investigation revealed that Reliant had manipulated natural gas prices, resulting in higher last-minute prices for gas needed to meet the state's power-peaking requirements. FERC believed it had insufficient evidence to proceed on most charges, but there was evidence that Reliant practiced coordinated trading on three occasions with British Petroleum (BP). For these violations BP agreed to pay $3 million in a settlement. The settlement also required Reliant to pay $25 million in cash, $15 million of which was due in thirty days. The settlement required Reliant to auction 824 Mw in California; the proceeds from those sales were to be placed in an escrow account up to a maximum of $25 million. That money was to be paid to the western states under established procedures. Pat Wood opined, "Today's settlement should serve as a warning to all energy companies that attempts to manipulate energy markets will have consequences." Reliant admitted no wrong doing; however, it did move to change some company operations and terminated the trader who had exercised the mischief. Reliant's chairman and CEO stated, "Reliant must assume responsibility for its actions. We intend to conduct our business in a fashion that not only meets the letter and spirit of the law but also sets a standard of excellence based on the highest ideals of corporate responsibility." However, California's attorney general continued to study the settlement agreement and withheld any comment. California spokesman Tom Dresslar said the settlement let Reliant off the hook and argued the company may have illegally withheld 15 percent of its generation capacity in the spring and summer of 2000. Dresslar commented, "While $50 million is nothing to sneeze at, there are some questions that need to be answered before a full evaluation of this order can be made." California subsequently complained about another FERC proposed settlement with Reliant involving a separate manipulation issue. FERC had been willing to accept penalties of $836,000 for conduct that produced $13 million of Reliant's profits. The issue involved allegations that Reliant sold electrical reserves for which they had been already paid. Had the ISO required that power to meet its peaking requirements, a number of blackouts could have been caused. This settlement did not relieve Reliant of its responsibility for refunding money for price gouging in California.

On October 9, U.S. Court of Appeals for the Ninth Circuit heard California's case against FERC in which the State had asked for $9 billion in refunds from the electrical energy suppliers. The three-judge panel was skeptical of FERC's contention that it acted legally three years ago when it did not require the suppliers to file the rates they would charge. The State argued that the Federal Power Act required the filing notices of specific rates, along with what they would charge. Such filings would have allowed the buyers to challenge rates and avoid market-based rates that had been allowed. FERC was constrained from relying on market-based rates without first getting Congress to change the statute. Judge Sidney Thomas questioned how a rate could be called unjust if the rate was not known.

On October 15, the House voted 229-182 approval of a resolution authored by Representative Lois Capps (D-CA) which rejected a compromise energy bill and which included authorization along with subsidies for oil and natural gas development in Alaska's Arctic Region. Fifty-five Republicans had voted for Capp's resolution. The House had previously voted in the spring against a proposed resources inventory. Critics had argued that the inventory proposal would end the congressional moratorium and executive orders that had prevented new leasing along most of the various states' coastlines for two decades. This included coastlines of the three western states in the moratorium; it was active until the year 2012. The oil and natural gas industries, along with manufacturers and the American Farm Bureau, urged the House members to reject Capps' resolution. The Farm Bureau maintained that if the moratorium were allowed to continue, it would result in economic disaster for American agriculture due to higher prices for fertilizers, agriculture chemicals, irrigation, and electrical energy.

On October 21, SMUD initiated the construction of the first phase of its Consumnes Power Plant, expecting to complete it in late summer of 2005. The plant was designed to generate 500 Mw of electrical power at a cost of $380 million. SMUD estimated the project could save $15 million to $25 million a year as opposed to buying replacement power on the wholesale market. In addition, SMUD had been seeking more natural gas to supplement the New Mexico gas fields that it had purchased in the spring for $135 million.

Natural gas prices spiked by 30 percent in the current month; PG&E's spokeswoman reported gas bills for the average household averaged $24.95 compared to $19.18 a year ago. PG&E declined to forecast prices, predicting supplies would be tight through the current winter and recommended conservation and investment in energy-efficient furnaces. Some experts believed prices would be relatively moderate. To the contrary, the DOE expected prices nationwide to increase by 9 percent. The CEC had predicted natural gas prices in California would be slightly higher that winter. Californians had been exposed to volatile gas prices after PG&E's prices in 2001 reached $125 a month for the average household. Meanwhile, state officials extracted $1.7 billion in a settlement with El Paso Corporation, which denied any wrongdoing. The state's natural gas supply had improved with the completion of pipeline expansion near Kern River, which would bring in additional supplies from the Rocky Mountain area.

Meanwhile, Governor-elect Arnold Schwarzenegger stood ready to develop his approach to the California energy situation. Schwarzenegger believed in free markets and wanted to give the energy free markets another try. Some consumer advocates, including Democratic lawmakers, believed his plan could result in another energy crisis. The governor-elect's energy policy statement stated that in order to assure low energy prices and to encourage new power plant construction, it would be necessary to make the energy markets workable. The Democratic legislature was not expected to embrace free markets, according to Joe Dunn (D-Garden Grove). However, there were many who believed California had to tackle the energy issue early on. Some of the State's long-term contracts would run out in a couple of years, and California still remained partially regulated. Many forecasters predicted California could run out of adequate power supplies by 2005, based on the assumption that the economy would recover and no new power plants would be built. The governor-elect planned to abolish the California Power Authority, as he believed it discouraged private power plant construction, and the independent suppliers had seen California as hostile territory. This was evidenced by Reliant's retirement of two of its small generating power plants. Their comment was, "It is not economical to invest in the required upgrades given the current market conditions in California and uncertainty regarding future market conditions."

Business lobbyists, however, didn't expect the governor-elect's energy plan to fade. California's Chamber of Commerce president Allen Zaremberg said, "He is concerned about job security and the economic climate in California, and energy is a big part of that. He'll go ahead and sign, [saying] 'This is going to solve the budget crisis in the long run if I can keep high-paying jobs here.'" The policy paper depended on creating enough power generation to prevent market mischief; it would require the utilities to maintain enough standby power to assure minimum reserve requirements to meet power demand shortages. Large consumers would be given reduced rates, provided they allowed the State to curtail their energy supply, if the state's supplies became tight. They would also be subjected to real-time pricing during peaking hours. Severin Borenstein opined, "This is a very progressive vision of how to implement market incentives in a way that could benefit customers." Additionally, Schwarzenegger would reestablish direct access for large consumers. He advocated California reducing its dependence on natural gas as a fuel for the generation of electrical energy. He favored alternative energy and renewable energy sources. Doug Heller, FTCR, predicted the governor-elect would abandon the State's $9 billion suit against the independent energy producers. Gubernatorial advisor Joe Rodota, disagreeing with Heller, said the governor would continue to pursue the refunds, and Ackerman concurred.

November

FERC on November 13 approved a $1.7 billion settlement between El Paso Corporation and the PUC over allegations that El Paso had deliberately restricted the flow of natural gas into California in 2000. The settlement required some changes that were still under review by El Paso and the California attorney general's office. El Paso, in agreeing to the settlement, admitted no wrongdoing. FERC's administrative judge found that the company had manipulated the market by withholding large amounts of capacity that could have permitted the flow of natural gas to California. The scheme had been accomplished through a subsidiary, El Paso Merchant. During periods of high electrical demand, much of the pipeline's capacity had gone unused, and gas prices in the state increased three-fold compared to the rest of the country. California charged in its suit with FERC that El Paso's scheming had cost the State $3.7 billion.

On November 13, Governor Davis signed an order terminating the energy state of emergency he had declared on January 17, 2001. Davis stated his energy policies had worked; touting that 9,500 Mw of new plants had been built since January of 2001. The rolling blackouts had ended in the spring of 2001, and electrical rates settled down the following summer. Many Democrats in mid-2001 had urged Davis to declare an end to the emergency, arguing he was jeopardizing the balance of power between the two branches of government. Gary Ackerman of WPTF observed, "California still sits on the precipice of future shortages. Thanks for the curtain call, but we'll keep slogging through this thing." Erik Saltmarsh, director of the state's EOB, commented that even though the immediate crisis had long since passed, lingering effects still remained. Saltmarsh went on to say, "We're still dealing with problems, and we have a lot of work to do to make sure [the crisis doesn't get repeated]." The EOB was still pursuing the $9 billion of refunds from suppliers for alleged overcharges, and the matter was still pending with FERC. A lot of consideration was being given to overhauling California's deregulation system. The Democrats wanted to return to a regulated system, while Schwarzenegger favored the free market approach.

On Wednesday, November 19, the U.S. Court of Appeals for the Ninth Circuit set aside a lower court ruling that would have allowed PG&E to spin off some of its valuable assets to new companies that would not be under the scrutiny of state regulators. The three to zero vote confirmed that PG&E did not have the right to automatically sidestep state law by creating new, less regulated companies in order to end its bankruptcy. The PUC was scheduled to meet on December 18 and make a decision on the proposed June settlement. The attorney general's office spokesman said, "To the extent that PG&E threatened to go back to its original plan . . . The Ninth Circuit today took that arrow out of its quiver. This is a major victory." PG&E's spokesman tried to shrug off the ruling and indicated it didn't affect PG&E's bargaining clout, as PG&E could still go back to the court and argue for its original plan. PUC commissioner Brown still wanted to see a settlement, but he was inclined to be less generous to PG&E and believed it need not happen by the end of the year. Commissioner Kennedy said, "This decision is absolutely critical to the economic health of the state. We have to have a solvent utility that can enter into contracts [and] build power plants. We could be facing another power shortage as early as 2006, maybe earlier . . . We can't afford to take time arguing over details." Schwarzenegger's press secretary said, "The governor understands how important it is for California's utilities to be creditworthy. He's looking forward to a fair but expeditious completion of PG&E's bankruptcy case. If it continues to drag on, it will delay necessary reforms that California needs to take to stabilize its energy markets."

The PUC unveiled three different proposals to end PG&E's bankruptcy protection and pave the way to final settlement. Such a settlement, it was felt, could affect PG&E's rates for the next nine years. Two of the proposals backed away from a previous preliminary agreement reached by PUC staff members and PG&E, which had been announced in June. The third proposal would have accepted the settlement in full. It was anticipated that some commissioners would offer their own proposals before the tentatively scheduled decision on December 18. Each of the proposed settlements would keep PG&E electrical rates relatively high. An estimated rate reduction of .5¢ would occur in January to be followed by another .5¢ reduction in 2008. Business interests warned that these high rates would impact Northern California's business climate and noted that, at the same time, SCE's consumer rates had started to drop. PG&E warned that unless the settlement was accepted, it would go back to court and seek to renew its original plan to transfer its hydroelectric facilities and other assets into new companies that would not be encumbered by state regulation. Under the current settlement, PG&E would remain regulated and would make various environmental guarantees. In exchange, PG&E would get $7.2 billion to pay off debts and would have a guaranteed after-tax return on investment of 11.2 percent, plus other guarantees that included dividend protection. The settlement was certain to provoke consumer groups' protests. Earl Bouse of CLECA said, "Here's the problem. The settlement is too rich for PG&E." PG&E responded that the settlement was the best way to end its bankruptcy case.

December

December found natural gas prices up by about 50 percent since the Thanksgiving holiday. The increases left the experts perplexed, as there appeared to be to be no fundamental reasons for those increases—no tight supplies or unusually cold weather. Executives of gas-dependent industries and the politicians alike wondered if the suppliers were back to their mischief of market manipulation. Analysts feared that these high prices could result in both higher heating and electrical costs for consumers, observing that natural gas had become the fuel of choice for generation of electrical energy. PG&E reported higher gas prices due to cooler weather; prices were up 4 percent since November, about 20 percent higher than the previous year. Natural gas prices on the futures market rose from $4.93 per MBTU at the end of November to nearly $7 in the week of December 8, and on December 12 the price increased to $7.22.

The U.S. natural gas industry had been experiencing a tumultuous period as the major petroleum companies backed away from domestic exploration and production of natural gas. Major companies like ChevronTexaco, BP, Royal Dutch/Shell, and ExxonMobil reduced domestic drilling in favor of international exploration in Indonesia, Niger, Qatar, and Russia. Their goal was to convert the gas to LNG for export to the U.S. This activity benefited smaller companies at the expense of the residential population and industrial natural gas users. Robert Ineson, director of Houston-based Cambridge Energy Research Associates, said, "That's what the future holds for natural gas." However, the problem with LNG was that there were only four LNG terminals in the U.S. at that time. Approximately thirty new terminals had been proposed and as a result of this new capacity, LNG imports were expected to quadruple by the end of the decade. Pipeline imports from Canada had been growing, but the Canadians were finding they needed their natural gas to satisfy their own requirements.

The Houston-based oil services firm of Baker Hughes reported on December 5 that the number of active drilling rigs was up 38 percent from the previous year, and that didn't account for the small drilling operators who were investing in older, less productive wells. The DOE had estimated the production growth at just 2 percent. But Lehman Brothers' analysis of forty-nine of the nation's largest companies reported the average daily production in the U.S. was down by 2.8 percent in 2003. Marshall Adkins, an analyst at Raymond James Associates in Houston, observed, "You wouldn't see as much pressure on prices if the major petroleum companies were more active." Still, he believed they wouldn't be able to reverse the trend.

During the preceding year, the major companies' domestic production had fallen dramatically. BP's domestic production fell by 48 percent, Royal Dutch/Shell's by 15 percent, ChevronTexaco's by 11 percent, and ExxonMobil's by 10 percent. Untapped domestic areas that the companies would have liked to pursue included the Rocky Mountain area, the coast of California, and the eastern Gulf coast, all of which were unavailable for ecological reasons. ExxonMobil was the sixteenth-most active driller in the U.S. The company's public affairs officer said, "If we could drill in certain areas of the U.S., we'd be there." Other smaller companies like

Apache, Anadarko Petroleum, Devon Energy, and EOG Resources actively acquired and drilled new wells.

Some traders and analysts suggested that some of the larger funds had bet at the beginning of the month that the price would fall below $5.00 per MBTU. There were others who bet on a scarcity of supply causing prices to rise. A Houston Citigroup analyst observed, "There's a great deal of speculative froth at the moment." Senate Judiciary Chairman Orrin Hatch (R-UT) planned to hold hearings after the first of the year to determine whether or not there was any market mischief occurring. Hatch stated, "Natural gas is so critical to U.S. consumers and the economy that if someone has been manipulating the market, they should go to jail. I feel we must determine once and for all if these price surges are the result of market forces or if there continues to be price manipulation."

At the time, natural gas inventories in the U.S. stood at approximately three trillion cubic feet, 7 percent above the previous year's level; according to the EIA, natural gas stocks were 3 percent higher than the five-year average. A spokesman for the CFTC confirmed that there had been increased surveillance of the large participants in the natural gas markets. This had become standard procedure following the wide price swings that had occurred earlier. Executives of the New York Mercantile Exchange reported they had not seen any evidence of market manipulation. They went on to say that the recent increases in gas prices were the result of reduced investments in the infrastructure, drilling activity, and trading activities following Enron's collapse two years previously. Drilling had increased in the Rocky Mountain area and in Texas, but these were fast-depleting wells. Additional imports from Canada had decreased due to Canada's own increased demand. Earlier in the year, natural gas shortages resulted in a number of layoffs and closings at ammonia, chemical, and fertilizer plants. Greg Lebedev, president of the American Chemistry Council, commented, "What's happening in natural gas markets has the very real potential to stop manufacturing in its tracks. At the moment, the price of natural gas is serving as the single largest brake on the U.S. economy." Federal Reserve chairman Alan Greenspan called for increased imports of LNG to avoid any attenuating effect on the economy.

Also in December, PG&E and TURN reached a new compromise plan for removing PG&E from the bankruptcy court. PUC commissioner Susan Kennedy said, "It's a huge breakthrough" and predicted it would be supported by a majority of the PUC commissioners. As had been anticipated, PUC president Lynch opposed the plan. The proposed settlement would reduce PG&E's average electrical rate 7 percent, from 13.8¢ per kwh to 12.9¢ per kwh in 2004, reducing the cost to the consumer by $1 billion over nine to ten years, about $1 million a year. Additionally, PG&E would forgo its corporate bankruptcy legal costs. A regulatory analyst felt that the compromise favored PG&E; it unduly allowed PG&E to set aside stock dividends of up to $900 million as opposed the pre-bankruptcy level of $400 to $500 million. Department of Finance's Dennis Miller didn't believe any agreement would have much of an impact on the state's economy and was quoted as saying, "When you get down to it, it's a pretty small piece of the pie. With the state's economy running at around $1 trillion a year, even $1 billion in electric rates pales by comparison."

December 12, federal bankruptcy judge Dennis Montali upheld the third proposed settlement that PG&E and the PUC staff had agreed upon in November, but he would not sign off on it until the PUC approved the settlement. The settlement would also require approval by the PG&E board and its parent company, PG&E Corporation. PG&E issued a statement to the effect that the commission's approval "is the only thing standing between our customers receiving immediate, significant rate reductions estimated at least $670 million starting in January 2004." It went on to say the plan would allow PG&E to emerge from the bankruptcy court as a financially healthy, investment-grade company. The official spokesman for the PG&E's creditors committee, Ed Feo, observed that the judge's ruling wasn't surprising and said, "We represent a lot of people who would very much like to be paid back soon." He also acknowledged that obtaining PUC approval could be a tough hurdle. The PUC was scheduled to vote its approval on December 18, and the approval did occur on that day.

The concept of wind energy took a hit this month. Environmentalists had originally supported wind energy, considering it an energy panacea—the answer to energy crisis. Environmentalists revisited that support after discovering that an estimated 22,000 birds had been killed by flying into the 7,000 spinning wind turbine blades within the fifty-thousand-acre Altamont Pass area, the largest wind farm in the world. Originally, Friends of the Earth had complained only about the ugly towers that ruined the view of the hills, but that was nothing compared to the bird kills. An estimated four to eight hundred golden eagles, along with thousands of red-tailed hawks, kestrels, and other raptors had met their deaths due to the wind turbines. Environmentalists sought to stop renewal of ten-year permits until a solution could be found to reduce the number of bird kills. Michael Boyd, president of Californians for Renewable Energy, said, "Renewing these permits without addressing the cumulative impacts on migratory birds, especially raptor series, will give a black eye to wind power." That organization, joined by the Center for Biology Diversity, asked the Alameda Board of Supervisors to stay a recent decision by a local zoning board that had granted renewal for some of the permits. Altamont Pass is an area considered to contain one of the largest populations of breeding golden eagles in the world. Those large raptors along with the thousands of red-tailed hawks use the pass in their route to their breeding grounds in the Central Valley. Of the sixteen other wind farms in the U.S., none comes close to Altamont's number of bird kills. The industry considers Altamont Pass as an "anomaly" providing many lessons for other wind farms, including Southern California's Tehachapi area and San Gorgonio Pass. FPL, owner of the largest number of wind turbines in Altamont Pass, proposed replacing 644 turbines with ninety-two larger turbines along with relocating the larger turbines into wind corridors with low rates of bird kills. FPL's re-powering plans, including four other companies, had been approved by Alameda County after completion of an extensive environmental review. However, the PG&E's bankruptcy case had caused enthusiasm for this re-powering to wane.

VII. As Night Passes—Year 2004

January

Wind power at this time accounted for 1 percent of the nation's electrical energy. Congress's procrastination on an energy bill was crippling the wind energy industry, as the economics had relied on tax credits that expired in December 2003. A renewal of tax credits was necessary to revitalize the wind energy industry. According to the American Wind Energy Association (AWEA), approximately $2 billion in economic activity had been lying dormant with the loss thousands of jobs. When it became evident the tax credit would be stalled, the industry started scaling back. Texas-based Cielo Wind Power idled $130 million of planned development that would have provided about three hundred construction jobs through subcontractors. Dallas-based Trinity Industries, which built wind turbine towers, had eliminated dozens of jobs and restructured its manufacturing facility toward other projects. Texas-based Lone Star Transportation transported giant wind towers across the nation but lost revenue and idled equipment as a result of the industry's gridlock. Lone Star vice president David Freebee observed that wind projects had offered some of the best returns for his company. He warned that lead times for wind projects meant much of the year would be lost once the tax credit expired. "If they pass the PTC [Production Tax Credit] today, that doesn't mean wind power will start tomorrow—it may not start again for another six months." He emphasized that the industry needed more consistent policies in order to do more long-term planning. It had been over a decade since Congress first enacted the wind tax credit legislation; the objective had been to motivate the infant wind industry, along with other renewable sources of energy. The industry stalled in 1999 with the expiration of the tax credits. The credits were renewed but expired again in 2001 and were renewed yet again in March 2002. The tax credit had already achieved its purpose, as costs had fallen steadily as a result of the increased activity. Some officials and legislators called for longer extensions of the tax credit, suggesting up to ten years, thereby giving the industry more time to plan. Trinity's senior vice president Mark Stiles observed, "If people spend the kind of money they spent in two to three tax credits, I can only imagine what would happen if the program were in place and had some stability." In states such as Texas with strong breezes and available vacant land, wind energy became cheaper than natural gas, and gas prices were expected to remain high. The one impediment to wind energy that remained was electrical transmission facilities—getting the energy to the consumer.

Altamont Pass had 5,400 turbines capable of producing 580 Mw—the average size of the turbines was 107 kw. The operating companies wanted to expand their generating capability. However, Alameda County imposed a moratorium on the construction of more wind turbines until there was a program to reduce the number of bird kills. The CEC concluded a four-year study on bird kills on Altamont Pass, finding the bird kill due to the wind turbine blades on Altamont Pass was about 1,766 to 4,721 birds every year; of those, 881 to 1,300 were hawk and eagle raptors. Recommendations had included replacement of the older turbines with newer, higher-rated turbines with ten times the capacity of the older ones. However, the replacement wouldn't be cost effective, unless Congress moved forward on federal tax credits that had been languishing. The new turbines would be required to be constructed as high as possible, but it was acknowledged that could pose a hazard to high-flying birds. However, it was believed fewer birds would be killed with larger blade diameter. One operator started painting a pattern on the blades in an attempt to make them more visible to the flying birds.

Senator Feinstein, concerned about another energy shortage in the summer of 2004, wrote a letter dated January 8 to Governor Schwarzenegger calling for immediate measures to control and expedite construction of new generating plants. An October report by the ISO had provoked her concerns and led her to believe that another energy crisis could occur within several months to three years. The report had projected that "adequate supply will most likely be available to meet the peak demands for the next five years . . . [but] reserve shortages could return as early as the summer 2004 under a scenario of high demand and adverse available capacity." The senator noted that private operators, as a result of deregulation, currently owned a significant share of the state's generating capacity, and many of those plants were being retired because of operating inefficiency; as a result, California was at risk of losing more than 7,500 Mw over the next several years. In addition, the construction of new power plants had stalled; less than a third of the permitted plants were under construction. ISO spokeswoman Stephanie McCorkle said crunch time was just two years away. The senator recommended that retail rates should be based on costs rather than the three-year rate schedule process currently used by the PUC and that steps should be taken to assure the rate structure was flexible enough to respond to the next crisis. She recommended that (1) investor-owned utilities be forced to plan and construct more than enough generating capacity in order to meet demand forecasts and to insure investments in transmission facilities and (2) that they should accelerate alternative energy and cogeneration planning and construction.

PUC's January 22 meeting had resulted in some decisions: (1) the utilities had four years to build reserves to between 15 percent and 17 percent, (2) independent power producers would be required to meet the same reserve requirement, (3) utilities were to have 90 percent of their summer energy requirements locked up one year in advance, and (4) they must meet those reserves as much as possible through energy-efficiency programs and renewable energy before building new conventional power plants or transmission lines.

Power generators and the ISO had been pushing for a 2006 reserve deadline to commence construction of new plants, arguing it would provide better protection against blackouts. Jan Smutney-Jones commented, "The commission today basically sowed the seeds for the next energy crisis." PUC president Michael Peevey said any new energy crisis "is unlikely and, if we do our job, it's not going to happen." Contrary to that opinion, the ISO continued to warn of potential supply problems. The commission required private suppliers to meet a 15 – 17 percent reserve requirement, and that set the stage for potential legal battles down the road. The private companies did not want the commission exerting any additional authority over them. Gary Ackerman of WPTF predicted his members would take the issue to the Supreme Court if necessary.

February

The PUC on February 26 accepted a formula for the distribution of the rate reduction it had approved in December. PG&E was ready to put the new rates into effect on March 1. As a result of the new formula, residential customers would have realized a 4 percent reduction, and large industrial and agriculture customers would have seen about a 15 percent reduction. PG&E spokeswoman Kristy Dennis said, "This is a new day. We're taking one more step in coming out of California's energy crisis." Bob Finkelstein of TURN said, "This decrease, although welcome, doesn't come close to undoing the damage done to consumers by deregulation." TURN was among the many groups that had previously agreed to the formula. Without TURN's approval there could have been months of legal battles among consumer representatives over how the $800 million rate reduction should be distributed. Commissioner Carl Wood requested the vote be delayed for another three weeks; he and Loretta Lynch both voted no to the approval. Lynch said, "I find little to be self-congratulatory about; we're still leaving significant money on the table." Overall, PG&E system-wide average rates would drop from 13.9¢ per kwh to 12.8¢ kwh.

March

On March 26, the PUC's Strategic Planning Division issued a report to the legislature recommending California wait until 2009 before attempting deregulation again. They believed that all of California's customers who had endured the 2000-2001 energy crisis should, in all fairness, keep sharing the ongoing energy costs, including paying off special bonds, assisting PG&E in emerging from bankruptcy, making continuing above-market payments to several independent power suppliers, and contributing to the long-term, above-market contracts to which the DWR was committed. PUC president Michael Peevey in a letter called the report "overly timid" and "a disappointment." Bob Finkelstein of TURN responded, "Rarely have I seen such a clear statement from a regulator that he's made up his mind and damn the facts, or the absence of any facts." However, the PUC position drew a rousing "Good for him!" from Jan Smutney-Jones of

IEPA, who countered, "There's no doubt that there's work to be done, but we shouldn't have to wait until 2009 to get things going."

In Smutney-Jones's view, the vital issue was how California's largest industries would purchase their electrical energy—whether or not it would be with the purchase options they had sought without putting a burden of higher costs and reliability on the other customers. In addition, the utilities had a problem with the design of the plan. If the utilities signed contracts and/or built power plants for their current customers, some of them might eventually switch to direct access and leave the remaining customers to pay for the new excess supply. The question was raised, who is going to build the power plants to meet the growing demand when they are not required until 2010? Gino DiCaro, CMTA, remarked, "A competitive energy market is probably one of the most significant ways the industrial community can reduce their costs these days." He went on to say, "Industrial customers saw their rates increase between 50 and 150 percent two years ago." Since then, rates had come down only 14 percent. DiCaro continued, "It's obvious a competitive market will drive those costs down." The Planning Division said that was not obvious at all. There was some speculation that legislation for a plan might be passed that year, while others thought the legislation would not materialize until 2005.

When federal bankruptcy judge Dennis Montali approved PG&E's $12 billion debt repayment plan in December of 2003, it set in motion the conclusion of PG&E's bankruptcy. On March 26, PG&E announced the conclusion would officially take effect April 12, 2004. Accompanying the action was a rate reduction for the company in March. UCEI's Borenstein observed that as a result of the plan, "everyone agreed ratepayers would pay lots of money until [PG&E] could pay off its creditors—that's how we're going to pay off the losses of the energy crisis." PG&E's consumers felt the effects. Under the terms of the bankruptcy reorganization, PG&E customers would be paying off obligations incurred during the crisis, and the bulk of that money would be going to the wholesale suppliers. Consumers were now paying substantially more than they had paid before the energy crisis.

April

With summer approaching, the ISO predicted no blackouts; it believed supplies would be "adequate if not ample." The ISO expected a new daily peak demand of 44,422 Mw during the summer, with the peak occurring probably between the hours of 4 p.m. and 6 p.m. However, if the unexpected occurred, the ISO would have to call for the consumer to start conserving again. The ISO observed that old power plants were being retired faster than new plants were coming on line, resulting in a net reduction in generation. Added to that was the fact that California's economy had been growing and the population had been increasing with an accompanying increase in energy demand.

On April 6, SCE filed a proposal with the PUC that would offer discount rates to large consumers, primarily manufacturers that were struggling to remain in business and were considering leaving California because of the high-energy costs. PG&E planned to propose a similar plan to the PUC in June. The California Chamber of Commerce's spokesperson opined, "The utilities have felt threatened by direct access; Edison is using a market-based, instead of a legislative-based, mechanism to respond to the threat." SCE had a plan that would have required interested businesses to file a financial-stress affidavit in order to obtain a five-year rate reduction. The initial discount would be 25 percent, followed by a 5 percent discount for each of the following four years. The plan would apply to businesses that used 200 Mw or more at any given time. SCE's Bill Bryan said, "Our rates are still higher than the national average. Our interest is in helping the state retain and grow jobs." Loretta Lynch believed the commission would have to look very hard at the proposal and expressed concern that small businesses and consumers would bear the costs of rate discounts applied to larger companies. Consumer advocate Bob Finkelstein argued that by giving discounts to major consumers, the utilities ran the risk of raising energy rates for smaller consumers. He added that smaller consumers would also face higher energy costs if large industrial users left the utility system. If large industrial consumers had discounts and direct access, retail consumers could be forced to pay more. "I'd be more comfortable if Edison said to the legislature, 'If you get rid of direct access, we'll offer these reduced rates instead.'" Chief economist Jack Kyser with the Los Angeles County Economic Development Corporation said, "There are plenty of other [out-of-state] locations that offer something for Southern California's manufacturing base." He added the proposed plan was "one breath of fresh air," as it had been very tough to do business in California. UCLA's Anderson Forecast had reported in

March that California in the last three years had lost more than 300,000 jobs and forecasted that the state would lose 1.2 percent more in 2004. SCE disclosed a net loss of eight large industrial users in 2002, while in the five years prior to 2002 it had gained an average of fifty-one manufacturing accounts annually. SCE also reported that large manufacturing companies using at least 500 kw annually had an average electrical bill of $750,000, and smaller companies using 200 to 500 kw paid an average of $141,000. SMUD's manager of major accounts reported that the company hadn't seen any loss in its commercial base and already had a rate-reduction program for large power consumers.

Governor Schwarzenegger, in a letter to PUC president Michael Peevey, said he supported a core/non-core market structure. "Countless businesses have told me that one of the greatest barriers to doing business in California is high energy prices. By fostering competitive wholesale and retail electricity markets that are properly monitored by regulators, California can begin to lower electricity bills and once again become the job-creation machine it once was." Consumer advocate groups warned that a core/non-core system could leave the utilities and their customers vulnerable to artificial shortages and price spikes like those the state had experienced in 2000-2001. Schwarzenegger's advisors indicated the governor intended to try limiting the lawmakers' role in redesigning the energy policy and would rely on the regulators whenever possible. The governor was also concerned that a complete rewrite of the energy policy would take several years and might cause delays in power plant construction.

Schwarzenegger indicated that he intended to take a two-pronged approach to California's energy problems by making a big commitment to renewable energy and by giving staunch support to the state's anti-global warming bill. Environmental groups honored Davis for having signed into law a number of environmental bills. It made Davis recall an incident when he was Jerry Brown's chief of staff. They were able to pass a law which they believed was very important and for which they received considerable press. Davis later called the head of the agency to see how the bill's implementation was proceeding. The head of the agency asked, "What bill?" Some environmentalist lauded Davis for having lobbied and signed into law the most expansive renewable energy requirement in the nation to date. However, the implementation of that law, it seems, was mired in bureaucracy.

It was observed that energy demand was coming back "big time," as demonstrated by the prices at the gasoline pumps and a Stage 1 power alert. It appeared the politicians and the media had lost interest in the energy issue when the 2001 energy crisis eased. With the energy situation easing and energy prices falling, power plant builders took a second look at natural gas, which remained California's principal fuel for electrical generation. An almost mirror image of the 2001 energy crisis developed as a result of limited supply. The City of Vallejo rejected a proposed LNG facility at Mare Island because of safety concerns. And there were worries that if California became deluged with imported gas, it might diminish incentives for renewable energy. State finance director Donna Arduin observed, "There is tremendous overlap with the energy agencies. We need to reorganize all this and make it much more efficient." She was talking about the collection of agencies that included the PUC, CEC, DWR, and the new Power Authority.

May

On May 21, a three-judge panel from the First District Court of Appeal in San Francisco issued a unanimous decision stating that the PUC could investigate whether or not a utility's parent company had failed in its obligation to keep its utility properly functioning during the 2000-2001 energy crisis. A PUC investigation had been initiated in 2001 over concerns that the utilities had transferred billions of dollars to their holding companies. PG&E's spokesperson called the decision "procedural," saying, "We continue to believe that the corporation and the utility have complied with all the applicable statues, CPUC decisions, rules, and orders." The state attorney general, however, filed separate litigation seeking restitution of $5 billion from PG&E along with at least $500 million in civil penalties.

A very interesting article appeared in the *Sacramento Bee*, May 24, regarding solar shingles. Premier Homes, a developer, was building a housing development in a Sacramento suburb and equipping the homes with solar shingles. The cost of the homes ranged from $270,000 to $400,000. The added cost for solar shingles was about $20,000. Apparently the added cost was not discouraging buyers, as there had been 400 applicants for the ninety-five homes. Senator Kevin Murray (D-Los Angeles), along with many of his Democratic colleagues, wanted to make solar equipment a mandated standard feature on all new homes, and Murray had been pushing a bill to do just that. In addition, Governor Schwarzenegger also wanted more builders to include solar equipment. Murray's bill had passed the senate and if the assembly approved it, the governor was sure to sign it.

Cities had looked at their energy options as the energy crisis developed in 2000-2001. A new form of municipal competition showed up for the three major utilities (PG&E, SCE, and SDG&E). Instead of forming municipal utilities for the takeover of the cities' electrical systems, five cities—San Marcos, Moreno Valley, Hercules, Victorville, and City of Industry—formed their own utilities to acquire newly built developer electrical systems. State law allowed cities and counties to assign to themselves the rights to utility infrastructures—sewer, water, electrical, and natural gas—that housing and commercial developers build. And under federal law, the private utilities cannot refuse connection to the grid by competing companies. Apparently the three utilities were attempting to combat those cities' plans to become municipal utilities through voter initiatives.

The U.S. Department of Energy reported that there was nearly 1.6 trillion cubic feet of natural gas in storage at the end of May. This represented about a five-year average and nearly a third more than the previous year's storage. The industry expressed confidence that storage through the summer would approach the 3.2 trillion cubic feet needed in the coming winter. Prices, however, were expected to stay above $6 per thousand cubic feet. The Natural Gas Supply Association (NGSA) reported a 26 percent increase in the number of rigs drilling for gas in the spring, but production had been flat due to declines in the production of older wells. A study commissioned by the NGSA reported that overall demand was expected to increase by as much as 3 percent to 51 billion cubic feet per day, with much of the increase coming from the electric power industry.

June

Tapes of conversations between Enron's day traders were released, and some were quite revealing. Selected samples appear below:

Sample One:

ENRON 1: "Well, he makes—ah, actually he makes between one and two a day, um, which never shows up on any curve shift, where he just buys it from a day ahead. He just f----s California."

ENRON 2: "Wait a minute, OK."

ENRON 1: "He steals money from California to the tune of about a million."

ENRON 2: "Will you rephrase that?"

ENRON 1 "OK, he, um—he arbitrages the California market to the tune of a million bucks or two a day."

Sample Two:

ENRON 3: "Um, if you took down the steamer, how long would it take to get it back up?"

ENRON 4: "Oh, it's not something you want to just be turning off and on every hour. Let's put it that way."

ENRON 3: "If we shut it down, could you bring it back in three—three to four hours, something like that?"

ENRON 4: "Oh, yeah."

ENRON 3: "Well, why don't you just go ahead and shut her down, then, if that's OK."

Sample Three:

ENRON 5: "All the money you guys stole from those poor grandmothers in California?"

ENRON 6: "Yeah, Grandma Millie, man. But she's the one who couldn't figure out how to f---- vote on the butterfly ballot."

ENRON 5: "Yeah, now she wants her money back for all the power you've charged right up, jammed right up her a---- for f---- $250 a megawatt hour."

Senator Maria Cantwell (D-WA) suggested Enron generated $1.14 billion in illegal profits by "gaming" the market, with the games facetiously referred to as "Russian Roulette," "Ping Pong," and "Death Star." She also stated that FERC had "fallen down on the job."

In Washington State, Snohomish County Public Utility District spent $100,000 to have Enron traders' tapes transcribed to paper. The tapes revealed how Enron had gouged western customers to the tune of $1.1 billion during the energy crunch of 2000-2001. The utility district had obtained the tapes over FERC's objections. An attorney for Nevada Power said, "Our hats go off to Snohomish. They have done a superb job of bringing to light the most outrageous conduct imaginable by an entity that was supposed to be regulated by the federal government." Snohomish had for months resisted buying any more energy while it waited for FERC to cap prices or otherwise control prices. Snohomish, Nevada Power, and others had tried to convince FERC that Enron should be ordered to return up to $2 billion in profits, but FERC did nothing on the case since it was introduced in October 2003.

A FERC ruling said Enron, Mirant, and Williams owed the DWR refunds that totaled $279 million. In 2003, FERC had ruled that sellers of energy had exceeded a market threshold and as a consequence, the sellers owed the State of California $3 billion in refunds. The fairness threshold should also apply to sellers who were billed for electricity billed during real time. FERC's action brought a letter response from Senator Feinstein on June 16. "FERC's inaction is deeply troubling to me in light of all the transcripts that have been released. Week after week we have seen evidence of massive manipulation and fraud perpetrated on the western energy markets," she wrote. "And yet FERC continues to fail to aggressively and sufficiently punish these perpetrators." FERC announced on June 17 that it would review the tapes. The Department of Justice and other agencies hadn't reviewed the tapes, believing a review would be too expensive.

On June 22, a federal appeals court rejected FERC's attempt to wrench control of California's ISO from the State.

June 24, 2004, there was an advertisement in the *Sacramento Bee* asking the public to support Assembly bill AB 2006. The advertisement, which also appeared on television, was sponsored by an organization named "Californians for Reliable Electric Service." The advertisements, it turned out, had been primarily funded by SCE. AB 2006 was a bill written by Assemblyman Fabian Nûñez (D-Los Angeles) in an attempt to correct the California electrical market.

California requested FERC to postpone further investigation into the refunds sought from generators in order to explore the possibility of settlements. California had been demanding $9 billion. On June 30, talks opened in commission chambers, and Governor Schwarzenegger read a letter that said, "Now is the time to resolve these cases and help put California back on the path to prosperity." A commission hearing was scheduled for July 5 to determine the amounts the generating companies should refund to California. Tom Dresslar, spokesman for the California Department of Justice, said many of the generators might now come to terms with the State, and some of those deals could reach hundreds of millions of dollars.

July

Enron's former CEO Kenneth Lay was indicted on July 8 for participation in Enron's energy schemes. In the early '90s, Enron had lobbied California in favor of deregulation and warned that the state's economy would become stagnant if it remained with the traditional regulated systems. In the crisis years, Lay had urged California to stay the course of deregulation. It's noteworthy that then-Governor Davis relied on Lay for consultation during the crisis years. In an interview with the *Sacramento Bee* at that time, Lay said, "Price signals need to be allowed . . . for people to understand there is a shortage."

Duke Energy on July 13 agreed to refund $207.5 million to California and other claimants. It was the fourth-largest energy settlement to date. Total refunds to date had been $2.63 billion. California had been seeking $9 billion in refunds, but it was unlikely to achieve that amount because of FERC's rulings. Erik Saltmarsh, executive director of EOB said, "FERC did everything they could to make that an uphill case for us. This is a good settlement." Duke admitted no wrong and stated, "We operated appropriately and within the rules of California." FERC allowed refunds to be made on a "no-fault" basis.

On Monday, July 19, at 4:26 p.m., California's combined three major utilities' electrical demand reached 44,042 Mw. That did not include the smaller utilities, munies such as SMUD, and cooperatives. The demand was not dictated by the weather but was the result of a growing economy. Consumption for the year had grown at 4 percent. The ISO reported that the demand did have a 7 percent reserve all day. Gary Ackerman of WPTF opined that the ISO's nervousness was "not misplaced." He said "We really don't have excess supply; we're dancing at the margin." Dynergy had recently told the *Sacramento Bee* that it was contemplating reviving two mothballed Southern California plants. They, however, still wanted to know how the State intended to design their energy market—"We're going to need to see some regulatory certainty."

On July 22, the state survived a fourth day of high electrical demand—even though PG&E had shut down one 1,100 Mw unit at Diablo Canyon and had additionally lost several other small plants. The demand that day hit 42,291 Mw at 4:20 p.m. Jim Detmers, acting ISO operating chief officer, said, "Things are going to get tighter and tighter through the summer." Another ISO spokesperson said, "The appetite for electricity has grown enormously over the past year." The ISO had predicted electrical demand to grow 3.5 percent to 4 percent during the current year; however, the growth had been actually 6 percent, according to the spokesperson. Dynergy's CEO cautioned that 2005 was going to get worse and 2006 even more so. He observed that lawmakers and regulators, to date, had failed to develop a new market design, which had left the investors nervous about putting any more money into power production. In addition, thirteen plants had either been put on hold or cancelled entirely for lack of investment. Calpine's spokesperson said, "We need some certainty in the market." In the meantime, AB 2006 was floating around in the legislature. In the generators' view, that bill would have given the utilities too much of a guarantee that they could recover their investment costs and would have encouraged the utilities to build power plants themselves—in effect locking the generators out of the market.

There has been considerable debate regarding the construction of LNG ports along the West Coast. In the contiguous forty-eight states there were only four LNG ports, and all were on the East Coast. No terminals had been approved for the West Coast. California LNG Project Corporation, a subsidiary of Mitsubishi Corporation, wanted to build the West Coast's first LNG port at Long Beach, and seven other ports were proposed for the region: three in Washington State, three in California, and one in Mexico. Opponents and environmentalists argued that the Long Beach port posed potential hazards to the city, as it could be subject to terrorist activity or earthquakes. They argued that the LNG facilities could produce a severe fireball that would destroy the port and buildings and scorch people as far away as one mile. James Fay, professor emeritus at the Massachusetts Institute of Technology and a LNG specialist, said, "If you're close when it happens and you are out in the open, you will be doomed. No one is going to [be able] come in and rescue those people." For its part, the LNG industry could claim an outstanding safety record and cited tight security, strong stable structures, and government safeguards. But protests killed terminal LNG proposals in the cities of Vallejo, Eureka, and Malibu. The Sierra Club, along with twenty-six other environmental groups, contested the state's pursuit of new power plants and urged more energy efficiency instead.

August

Summer was in full swing, and experts modified the "Flex Your Power Now" program. They asked consumers to cut back on their energy usage between the hours of 4 p.m. and 6 p.m.—the usual peak demand period for each day. If the ISO anticipated an energy alert, a message would be sent out in the evening or in the morning before the alert was to occur. This emergency call was issued four times during the summer, and consumers reduced their power demand by 5,500 Mw.

On August 10, the state set a new record for a peak energy demand of 44,497 Mw and anticipated reaching 46,000 Mw the following day. Fortunately, the Pacific Northwest, according to the ISO, had megawatts to sell. Apparently the issues between the governor and the Democrats concerned who would be allowed to build new power plants and whether big consumers should be allowed to opt out of the utilities' service and contract directly with generators for their energy. The governor favored a bidding process for the construction of new plants. The Democrats feared that allowing generators to build power plants would risk another energy crisis. Governor Schwarzenegger opposed the Democrats' bill AB 2006; he wanted to return to a free unregulated market. Private generators complained they were unable to obtain financing because California did not have a firm market design policy.

September

California experienced record-high electrical energy levels during the summer of 2004 and anticipated needing additional generation capacity over the next two years to meet the demands of an increasing population and the growing economy. Electrical consumption was up by 6 percent. The state's population was increasing by half a million people annually, and 200,000 home-building permits were being issued each year. Some 8,000 Mw of new generation had been built since the 2001 peak energy demand, and construction of additional plants was under way. The CEC reported there were enough plants under construction to supply 3.5 million homes; however, there was 8,000 Mw of additional generation that was either on hold or scrapped for financial reasons. But it was questionable whether the state would see any increase in power plant construction. Power plant builders were plagued with financial problems and also had to contend with California policy makers' ineptitude to fashion a suitable energy market. John Olson, investment analyst with the Houston investment firm Sanders Morris Harris, said, "There is too much uncertainty." He opined investors were leery about investing in California, given the state's red tape and their concern over PG&E's bankruptcy. He believed investors were waiting to see the design of California's energy market. The state assembly had passed AB 2006, a bill that would have returned the state to a partially regulated market and allowed the utilities to again build power plants, including the proviso that the utilities could pass construction costs on to the ratepayer. Governor Schwarzenegger indicated he would veto the bill if it hit his desk. The veto would keep in place a two-year-old law that required utilities to hold an auction to give power plant builders the opportunity to compete in building power plants. If the governor vetoed AB 2006, consumer advocates threatened a ballot initiative.

Increases in natural gas prices had provoked new interest in coal-fired electrical generation. Coal is an abundant fuel throughout the planet, is plentiful in the U.S., and is cheaper than oil or natural gas. Currently about half of the electrical generation in the U.S. is obtained from coal-fired plants. Surveys by the DOE and Virginia-based Energy Ventures Analysis estimated that between ninety to one hundred coal-fired power plants had been proposed in the U.S. Joe Lucas, vice president of the Florida-based Center for Energy and Economic Development, opined, "There's been a recognition that you can't power America's energy future without coal." Vickie Patton, a senior attorney in the Rocky Mountain office of Environmental Defense, observes "We're seeing coal-plant proposals just mushroom across the interior west." Sempra Energy Resources had proposed building a large plant near the site of Burning Festival in Nevada along with one in Utah. Steve Ellenbecker, energy policy advisor to Wyoming's Governor Dave Freudenthal, considered Wyoming as home to resources of coal, oil, and wind and saw exploration of these resources to be the core of that state's economy. Developers had already approached the state with proposals for six to eight power plants. They were investigating options with other western states and were spending an estimated $4 to $6 billion on a transmission network to export energy to California.

Coal, unfortunately, is associated with a number of environmental scourges—acid rain, tainting of streams, haze, and production of CO_2. Environmentalists argued that the construction of coal-fired plants would commit the U.S. to fifty years of polluting technology that could be avoided with increased energy efficiency and renewable energy—which they see has a panacea to our energy needs. California, though the largest consumer of electrical energy in the western states, was adamantly opposed to constructing coal-fired power plants inside the state, but it hypocritically imports 20 percent of its electrical energy from coal-fired power plants located outside the state. The state had no regulations that prevented the construction of coal-fired plants, but any suggestion of building coal-fired plants within the state would be a political fireball. James Hahn, mayor of Los Angeles 2001 – 2005, had told the LADWP to cease its expansion of coal-fired plant expansion in Utah. V. John White of the Sacramento-based Center for Energy Efficiency and Renewable Technologies argued that the State had the clout to design regulations and rules to restrain construction of coal-fired plants outside of California. SCE and SDG&E wanted the ability to seek low-cost sources of electrical energy, including coal, but PG&E was hesitant to buy coal-fired energy, believing costs could increase by State-imposed CO_2 emission controls.

On September 9, the U.S. Court of Appeals for the Ninth Circuit gave California the opportunity to again pursue approximately $2.8 billion in refunds. Attorney General Lockyer believed the State could support arguments for an additional $3 billion. The ruling held that the FERC had the authority to order refunds. It also ordered FERC to initiate proceedings into how much additional money California was owed. However, if California prevailed, it would be years before the consumer realized any benefit from the refunds, and the biggest share would probably go to big businesses as they had paid the biggest share during the crisis. The electrical marketers argued that the decision opened the door for more turmoil, just when the state was in the need of more power plants. Jack Hawks of the Electric Power Supply Association said, "This [issue] has dragged on for four years now; the last thing California needs is more uncertainty." FERC did welcome one provision of the ruling that validated FERC's authority to establish market-based wholesale electrical rates as part of deregulation. The State had been arguing that FERC should have been much more aggressive in insisting power marketeers file detailed reports to ensure rates remained fair. The court held that FERC could not enforce fair rates without seeing detailed filings from the marketers, which it had not received. The court held that "despite the promise of truly competitive market-based rates, the California energy market was subjected to artificial manipulation on a massive scale. With FERC abdicating its regulatory responsibility, California consumers were subjected to a variety of market machinations." California had always argued it had been overcharged by at least $9 billion. However, FERC believed that the State could make a case for only $3 billion. FERC argued that $6 billion was disallowed because four months earlier in the crisis it had excluded certain kinds of trades from refunds. Yet California had not received any of the $3 billion FERC had considered refunding. The State had, however, negotiated approximately $800 million to $1 billion in settlements with various other marketers.

On September 25, Governor Schwarzenegger vetoed AB 2006, as he had promised to do for several months. He favored a less regulated open energy market and in particular, direct access for businesses, and AB 2006 would have eliminated that option.

October

The Dairy Power Production program, sponsored by the CEC, was created in 2001, and the California dairy industry entered the field of biogas electrical generating—energy being derived from burning the methane gas in cow manure. The California dairy cow population, at the time, was 1.7 million cows contained in 1,950 dairies that generated approximately fifteen million tons of manure solids each year. The state's fourteen pilot projects had been expected to generate 3.5 Mw. They had cost an average of $1.2 million each, with about half the money coming from state grants and other support from federal money. One dairy farmer's project was a 160 kw plant that cost $773,000 with 40 percent of that paid by the State. By the end of 2004, twelve of fourteen state-funded pilot projects were expected to be in operation, followed by three additional units in 2005.

However, this precarious biogas program was two years behind schedule. For one thing, the process had been under scrutiny regarding its environmental benefits. The objective was to reduce the manure odor, support alternative energy goals, and reduce the emissions of greenhouse gases. In addition, the program had been hobbled by technical difficulties, milk prices, state budget cuts, and it apparently required more utility co-operation. Lack of utility cooperation had become such a problem that a senate committee addressed it. Mark Moser, president of RCM Digester, which designed one dairy farmer's methane digester, commented, "[Utilities] just don't want to help—the more high-priced energy they sell, the better it is for them." Supporters of dairy power fought a battle over a package of proposed tariffs that had been proposed by the three major utilities. Environmentalists and farmers believed the utilities had purposely designed the tariffs to undermine biogas. A San Francisco environmental group named Sustainable Conservation, a supporter of dairy biogas power, accused the utilities of spending an average of one year to approve connection permits to their power systems. PG&E officials contended they supported the program and had been working to reduce the confusion that had slowed progress. A PG&E spokesperson observed that small independent producers created complications and even problems in rural areas. "A lot of times, folks look at this as plugging in a toaster, but it's very far from that." PG&E planned to meet with dairy interests in December to resolve their concerns. The Sierra Club opposed methane power plants, being scared that huge dairies would increase the cow population to an undesirable level. Additionally, they didn't consider that type of energy as "renewable," since it was the byproduct of an "inefficient wasteful industry."

State Treasurer Angelides announced on October 26 he was not convinced that the environmental benefits were there. He put a freeze on one state pollution prevention fund on the grounds the State might have to back away from projects that lack environmental benefits.

November

On November 29, federal authorities unsealed indictments that accused five former natural gas traders of reporting fake trades to industry publications. Three were former El Paso traders, one was a trader at Dynergy, and one was a former trader with Mirant. Five traders from El Paso had already pleaded guilty to reporting fake trades.

December

On December 14, Governor Schwarzenegger had energy on his mind. He visited the ISO in Folsom to celebrate the completion of the eighty-four-mile transmission line addition along Path 15. The project cost $250 million, to be paid for by ratepayers, and added an additional capacity of 1,500 Mw to Path 15 transmission facilities. The governor assured reporters California would not be faced with an energy crisis in the summer of 2005—"Trust me, everything will be under control—your lights will be on." The governor also signed an executive order that required all state buildings to be 20 percent more energy efficient in the next decade. He had his administration reviewing LNG technology to determine its risks and benefits. The governor opined that the key to preventing another crisis was to build more power plants, so he wanted to make the permitting process easier and less expensive so as to expedite the construction of new plants. Mindy Spatt of TURN observed that his language "is often a euphemism for less regulation. And to the extent we had a problem it was created by deregulation . . . Certainly we need more power for the future, but the problem is when you tie that to the fears that people have of blackouts—because the blackouts were caused not by lack of supply but by an excess of greed."

On December 16, Governor Schwarzenegger announced his selections to fill the two PUC commissioner spots that would open on January 1. His selections would replace chair Loretta Lynch and commissioner Carl Wood. The selected replacements were Steve Poizner, 47, a Republican of Los Gatos and Dian Grueneich, 52, an energy and environmental businesswoman. Both supported the direct access concept. The PUC's philosophy was expected to shift to harmonize with the governor's philosophy, embracing a free-market system and looking dubiously on organized labor.

VIII. Shades of Darkness—Review

Energy is the essence of California's economy. California's experience in the energy squeeze that emerged in the year 2000 demonstrated just how vulnerable the state was, and is, with regard to electrical energy. Three factors contributed to this vulnerability: (1) the politics and economics of the energy industry; (2) environmental concerns; and (3) the population's attitude, naïve at best and arrogant at worst, that their own state should not pollute itself with electrical energy generation but could and should exploit and consume the resources of other states, the environmental or economic needs of which were not a matter of concern. For all these reasons, California had had a tacit moratorium on the construction of power plants for ten to fifteen years prior to the crisis of 2000-2001.

In 1995, the PUC ordered investor-owned utilities to contract for energy with private companies that were planning to build 1,400 Mw of new plants. SCE objected and appealed to the FERC, arguing they would not need that power until 2005. More realistically, however, this reluctance was due to anticipated deregulation of the energy market. When the idea of deregulation began to materialize, the utilities quit building power plants because they figured they would be required to divest themselves of them. In 1996, the state did pass deregulation legislation, which slowed construction even more. FERC sided with the SCE, and those plants were never built.

In 2001, a review of twenty-one proposed power plants, including those under construction, was made and the review reported their progress was hampered for variety of reasons: environmentalists slowed or stopped the progress on some power plants, but so had labor unions, corporate neighbors and others that had abundant resources at their disposal. Competing power generators complained the loudest about environmental obstacles, a ruse; they were using those arguments to delay the licensing of competitors. Adding to the problem was limited natural gas resources along with limited pipeline capacity.

The CEC revealed in 2001, twenty-two energy companies had intervened in twelve of their competitors' proposals, for some the objective was to simply to ferret out competitor information and many were competing for the availability of natural gas and transmission capacity. David Holberg, civil engineer and writer for the Ayn Rand Institute, observed that green activists had worked for decades to stop the construction of major power plants in California, and observed they had succeeded. Environmentalists' concerns dominated the state's energy policy. The thought of nuclear power in California was nonexistent. He also observed that the recent blackouts evidenced that the activists goal is to protect fish, trees, water and at the expense of mankind however, the NIMBYs had their deft hand in there also

The state's inability to meet its electrical demand due to inadequate supply opened the door for energy suppliers to game the market. The situation was exacerbated by the regulators' (FERC and PUC) failure to meet their fiduciary duty to maintain "fair and reasonable" energy prices that would have allowed the utilities to recover their costs plus a reasonable return on their investment. FERC allowed the wholesale energy market to float, and float it did—at the utilities' and the State's expense. Wholesale rates went unrestricted while retail rates were totally restricted, pushing the utilities to the edge of bankruptcy. In complete contradiction to the objectives of the Federal Power Act passed in 1935, FERC sat on its hands and allowed the generators and marketers to devise their own wholesale energy prices at whatever level they could get away with. Governor Davis and his appointed PUC president Lynch were both steadfastly determined not to increase retail energy prices. The private utilities were required by law to supply electrical energy[14] to the consumers within their territories, and they had met that obligation. But to meet their obligation to serve, the private utilities ran out of money and took on enormous

---

[14] In January 2001, the PUC voted 2 to 1 to order the utilities to keep power flowing, citing the utilities' "obligation to serve."

debt, sending PG&E into Chapter 11 bankruptcy. And the removal of direct access financially impacted the large users of electrical energy.

The result—California for ten to fifteen years prior to the 2000-2001 crisis had been comfortably relying upon neighboring states for about 20 percent of its electrical energy. The public had been seemingly oblivious to the fact that someday those neighboring states might require that energy to meet their own needs and that their hydro resources, in particular, might not be dependable.

In the year 2000, that scenario actually occurred. That was the year of the drought in Oregon and Washington, and those were the very states that had been providing California with electrical energy from their hydroelectric power plants. The crisis was exacerbated when Arizona's and Nevada's expanding economies started demanding their own states' energy generation.

In January of 2001, California's crisis really began unfolding. The state's hydroelectric potential was diminishing due to reduced water supplies. The ISO reported problems with the importation of power. The Western Governors Association concluded the combined western states did not have enough power to satisfy their growing populations and economies. The energy situation was aggravated by the limited thermal capacity of Path 15, which transmitted power from Southern California to Northern California. Power companies had considered building 3,000 Mw of generation along Path 15 but became discouraged without an upgrade of the Path. By January 9, reserves were especially tight. The system lacked 10,000 Mw, the result of planned and forced outages. The ISO was forced to issue Stage 2 emergency alerts. January 1, 2001, the DWR was authorized to commence buying power for the state's three beleaguered utilities, which were seriously in debt and close to bankruptcy.

January 21, 2001, California experienced its first blackout. Consumer advocates believed the blackout was finance driven and maintained that outages would not have occurred had there been enough money on the table. With that mentality, Californians were obviously "in the dark" about the state's energy problem; they believed there was adequate generation available.[15]

---

[15] A poll in the *Los Angeles Times* indicated that more than 54 percent of Californians didn't believe there was a power problem. Had they been required to pay the appropriate rate for their energy, that poll would have

SDG&E consumers' response to a call for conservation resulted in their lowest demand since 1997. Governor Davis sought an extension of the DOE's order compelling energy suppliers to sell to California, an order scheduled to expire in February 2001. However, even with the energy crisis gaining momentum, the public remained preoccupied and generally oblivious to the state of affairs.

In March 2001, during the energy crisis, the Auburn Dam project once again surfaced. Thomas "Rico" Oller, a Republican representing California's 1st Senate District, introduced legislation asking for a statewide vote to issue State revenue bonds to construct the dam. In addition to power deficiencies, water shortage had become a prominent issue in California. There had been no new water storage capacity built in California since 1976. Demand for water had risen dramatically due to the state's uncontrolled population growth and new environmental restrictions. The Auburn Dam could have put many of those concerns to bed, generated electrical energy, and added a water reservoir.

Congress had originally authorized the Auburn Dam in 1965, and it had been expected to produce 300 Mw of electrical power. By 1967, roughly $300 million of federal money had been spent on the initial construction of the dam. However, earthquakes in 1971 and 1975 caused serious concerns. The public feared not so much that the dam would rupture, but that a failure of the Auburn Dam would in turn cause the failure of the Folsom Dam downstream from it, and waters would finally inundate the city of Sacramento in something like the 1889 Johnstown, Pennsylvania flood. Construction was halted in 1975, and the Auburn Dam project gave way to the idea of raising Folsom Dam by seven feet. But what was the purpose? Would a higher Folsom Dam allow greater water storage, or would the current storage level be maintained and the additional capacity be there, just in case, to catch the California "Johnstown flood?" The latter scenario was presumed. These concerns were resolved a few years later with the proposed raising of the Folsom Dam. However, by then the free-flowing American River had become an important recreational area. Environmentalists created a difficult opposition bloc that still exists. In 1992, the Democrats defeated the project by a substantial vote. In 1996, four years later, the Sacramento-area federal delegation proposed

---

been considerably different.

building a $950 million Auburn Dam, but the House Transportation and Infrastructure Committee killed the project.

The purpose of Oller's 2001 bill was to allow the state, rather than the feds, to finance the construction of the dam. The estimated price tag was $2.5 billion, to be recovered by selling power and water. However, first Congress had to approve a land transfer or lease of the dam site and lakebed property to allow the project to proceed under state authority. Unfortunately, the environmentalists, supported by the state legislature, again failed to consider the state's future needs and defeated the bill. However, the white-water enthusiasts may have lost their white-water sports. The general public generally does not look beyond their noses and their own selfish interests, but who can blame them? The state has no energy czar to guide them through these issues. On June 26, 2001, Congressman John Doolittle (R-CA) ended his quest for an Auburn Dam study, recognizing that he would have an extensive uphill fight in the House of Representatives

May 2001 demonstrated just how vulnerable California was. On May 7, an early heat wave coupled with several power plants down for repairs and/or upgrades caused a series of blackouts in the late afternoon. The ISO director of operations had predicted in an earlier press conference that the state would probably escape power interruptions for the day. But less than three hours later, the grid operators were surprised by blackouts. The ISO operations director observed that the load ended up 2,500 Mw higher than it had been when he walked in that morning. He predicted more blackouts would occur, as there had been more than a fourth of the state's generation off line. The situation was exacerbated by a continuing heat wave, which continued into May 9, with several Central Valley communities setting temperature records.

It was estimated that power demand in the summer would exceed 33,000 Mw. It was expected to be touch and go by late spring, with approximately 30 percent of California's generating plants down for annual maintenance and/or unplanned outages. California's neighboring states all experienced heat waves that summer and had power plant problems of their own, thereby cutting power available for export to California. Four western nuclear plants were down for refueling and maintenance. The DWR reported that BC Hydro had withheld promised power for several hours unless it received payment up front from California. A three-hour loss of imports from Canada brought the California grid to within five minutes of blackouts. The state then lost 1,000 Mw of power due to a transmission problem in western Canada, resulting in California blackouts lasting forty-five minutes to one hour.

During the crisis, CEO Keith Bailey of Williams suggested that FERC temporarily cap wholesale profits at 15 percent above the cost of production for the period between May 1, 2001, and the fall of 2002. His objective was to save California's energy market and encourage the building of new power plants—once generators were confident they would be paid for their energy. Californians then would be assured they were buying power at reasonable prices, and the suppliers would be satisfied in receiving a reasonable profit. The suggestion of a 15 percent profit was more than most suppliers sought, and it would have encouraged investment in new power plants. Had FERC implemented this suggestion, the mischief by the marketers would have disappeared. It was an excellent idea, which would have provided immediate results.

In 2001, the utilities and the State spent $347 million total on rebates and promotion of conservation, with costs divided between the ratepayer and the taxpayer. Those that enjoyed rebates made their contribution to the rebates themselves and were supported by all of the other ratepayers. The taxpayer and ratepayers paid for the promotions.

The CEC predicted the state would have adequate electricity in the year 2002, even if the summer was unusually hot. Predictions were based on the assumption that people would conserve half as much as they had in 2001. The CEC also anticipated being able to import between 1,000 to 3,000 Mw from the Northwest as a result of that region's heavy snowfall. Additionally, new power plants in the Southwest were scheduled to come on line and were expected to help the supply. However, July was expected to become uncomfortably close to excessive demand, and the possibility of plant shutdowns remained a big concern. The Consumers Services Department had spent about $62 million in the previous year reminding people to conserve and planned to spend about $57 million in its "Flex Your Power" advertising program in 2002. However, not apparent to the taxpayers was the fact they would be paying for those reminders.

Many were predicting California would once again be faced with a shortage of electrical energy by the years 2005 and 2006. If that happened, California figured it would again be looking to its neighbors to fill its electrical energy supply gap. Many producers like Calpine and Duke had put power plant construction on hold, mainly due to a combination of low electrical energy rates, relatively high natural gas prices, and the lack of a comprehensive energy policy. *Dèjá vue.*

California's demand in 2002 was 31,114 Mw, and its imports were 7,177 Mw—a 23 percent import requirement. That year, for California to be fully self reliant, it would have needed 7,177 Mw of additional generation just to play catch-up and another 4,667 Mw to provide a 15 percent reserve, or a total of approximately 12,000 Mw of new generation to become totally self-reliant with firm power. And at that time, a maximum demand of 44,000 Mw was projected for 2004—an additional 12,000 Mw. The question then was, how was the state going to accomplish that, particularly without a comprehensive wide-ranging energy policy?

In 2002, Calpine and Bechtel proposed a 600 Mw power plant to be built adjacent to PG&E's Metcalf Substation in south San Jose's Coyote Valley—an ideal location, as the plant could be connected directly to the transmission system. Cisco Systems, acting as a NIMBY, supported by San Jose's mayor and city council, contested the project. The NIMBYs frequently undermine practical solutions to the state's energy facilities.

Calpine, on January 16, 2002, brought the building of their proposed power plants to a halt, and that put billions of dollars of power plants nationwide on hold—including 4,800 Mw in California. Electrical energy supply became a nationwide problem. Calpine could proceed with construction only if it was able to obtain backing from finance institutions, institutions that required sufficient demand to be available at a suitable price. State officials were concerned that a slowdown in building plants could result in a repeat of the 2002-2001 crisis, as the economy had been recovering with a substantial increase in demand for electrical energy. David Freeman, chairman of the state's Power Authority, commented, "The power plants are dying like flies out there—how dumb can we get? It's pretty clear that shortages mean higher prices. These folks have every incentive in the world to keep us on short rations. As soon as prices come down, they lose interest in building power plants." Peter Navarro, professor at UC Irvine, observed, "The problem with deregulation is you get boom or bust." Other experts saw that California, for the foreseeable future, had entered into long-term contracts and believed there was little chance for a major shortage anytime soon. The DWR predicted the State would be buying and paying for large amounts of power and selling much of that power at discount prices, thereby costing the State billions of dollars. Severin Borenstein said "Companies like Calpine are merely putting on hold projects that would have made sense only if prices had stayed sky-high . . ." Calpine's shareholders and lenders, obviously, had to be concerned about their balance sheets, and they questioned the company's plans to double the kwh it produced in California. Once again, the vulnerability of electrical energy supply was demonstrated in that there is no assurance that a competitive energy market will provide adequate generation just to meet demand, let alone provide a reserve. What are the incentives needed to encourage the private sector to construct enough generation to meet the state's electrical energy requirements plus a reserve? It is very questionable that a competitive market will do it; generators and lending institutions need assurances that there will be an income stream to support power plant projects.

The 2002 diminishing power plant construction caused Steve Larsen, executive director of the CEC, to wonder, "At what point do companies decide prices are low enough that they don't want to build any more? I don't know how to forecast that." A FPL spokesperson commented that plants would not be built if they were not economically viable.

On July 31, 2002, FERC proposed a plan to standardize electricity trading and transmission access across the contiguous continental U.S. The plan would have given FERC increased control over pricing, planning, and operational decisions that once belonged to the individual states. The plan prompted considerable static from fifteen states. The individual states feared approval of the plan would force the states toward deregulation, driving up prices in those states where electrical energy was inexpensive.

The plan envisioned a national grid overseen by independent regional boards that would be in a position to determine when and where new power plants needed to be built, including the transmission facilities to service them. The transmission network was to be available to all generators under a single rate structure set by FERC. FERC said there would be newly-designed rules to prevent the kinds of games that allegedly helped drive up prices in California in 2000 and 2001 and that the plan would ensure adequate power would be always available to all consumers. FERC argued the new rules would encourage long-term contracts; minimizing potential price swings on the spot market. Here again, long-term contracts would lock in rates to buyers for years to come, unless, of course, escape clauses were needed to be included in the contracts and take-or-pay contracts should be avoided. FERC opined there would be regional transmission price differences designed to avoid transmission bottlenecks and congestion—that contention should be questioned. FERC believed the new rates would be designed to serve as incentives for investors to build new facilities. Again the question is *how*? The impact of the plan would be the greatest in those states that were still regulated, requiring them to surrender control over their transmission systems to the FERC's regional agencies. FERC, however, said it wanted to respect the needs of municipal utilities. FERC generally sits back and says, "Trust me." I, however, see it as an agency that is seeking greater power. Its performance during the crisis years was an impressive display of incompetence.

In December 2002, the media reported that the CEC and the PUC were collaborating on experimental programs to equip consumers with high-tech meters that would measure how much energy they were consuming and the time of day that the energy was used. The objective was to allow consumers to save money by rescheduling their energy requirements to avoid peak usage. The CEC, with PUC approval, planned to equip 2,500 volunteer small businesses and residences with the new high-tech meters. The volunteers were to be divided into three groups: (1) time-of-use, (2) critical-peak-pricing, and (3) utility-control.

The CEC and PUC estimated use of these pricing systems would result in shifting 10 percent of peak demand away from the normal peak demand of approximately 4,000 Mw—the equivalent of eight 500 Mw power plants. An interesting and worthwhile program if implemented—it would be interesting to know the results. Currently, rates are calculated at the end of the day; the energy used is divided by its total cost and then the consumer is billed based upon the average price. This average method basically subsidizes consumers who use energy during high peak periods by the consumers who use their energy during the low demand periods

The U.S. Senate voted on June 12, 2003, its approval to support a bill directing the Department of the Interior to study the extent of oil and natural gas reserves available off the West Coast, the Atlantic Coast and the Florida Gulf Coast. An amendment to block the exploration was initiated by Senators Feinstein and Bob Graham (D-OR), who along with others argued that the only reason for the exploration was to exploit those reserves. Senator Pete Domenici (R-NM), chairman of the Senate Energy and Natural Resources Committee, assured them that the provision would not lift the moratoriums; it would simply direct the secretary of the Interior Department to conduct a study. It is a prudent move to take an inventory of our energy resources and where they are located. Domenici wisely proclaimed, "Knowledge is better than no knowledge." Senator Mary Landrieu (D-LA) added, "This amendment doesn't have anything to do with drilling or lifting. It will simply produce information that will allow us to make informed decisions." The bill was defeated by a vote of fifty-four to forty-five. Senator Feinstein argued, "The real intent is clear. California knows what's at stake, and we have made an 'informed decision.' We do not want drilling off our coast." Ms. Feinstein, yours was not an informed decision—yours was a passionate decision. One day there will be drilling off California's coast despite Senator Feinstein; however, it may not be in our generation. It may be one hundred years or even one thousand years from now. Now is the time to determine what resources are available to us and plan for the future by making decisions on how best to utilize them. We owe it to future generations.

Drilling did increase in the Rocky Mountain area and in Texas, but that was on fast-depleting wells. Imports from Canada decreased due to Canada's own increased energy demand. Earlier in the year 2003, natural gas shortages resulted in a number of layoffs with the closings of ammonia, chemical, and fertilizer plants. Greg Lebedev, president of the American Chemistry Council, commented, "What's happening in natural gas markets has the very real potential to stop manufacturing in its tracks. At the moment, the price of natural gas is serving as the single largest brake on the U.S. economy." Federal Reserve chairman Alan Greenspan called for increased imports of LNG to avoid any attenuating effect on the economy. Once again, do I have to remind the public that *energy is the essence of our economies*? Where is our comprehensive national energy policy?

On March 15, 2004, the PUC's Strategic Planning Division issued a report to the legislature advising that California should wait until 2009 before trying deregulation again. They postulated all of California's customers who had endured the 2000-2001 energy crisis should, in all fairness, keep sharing the ongoing long-term contract costs, including paying off special bonds, assisting PG&E in emerging from bankruptcy; making the continuing above-market payments to several independent power suppliers, and paying for long-term, above-market contracts the DWR had committed to. A vital issue was how California's largest industries would purchase their electrical energy and whether it would include the purchase options they sought without putting a burden of higher costs and reliability on the remaining utility customers. The plan would have the utilities sign purchase contracts and/or building power plants for their current customers. The plan must consider those customers who may eventually move to direct-access and the impact on the remaining utility customers who would be left to pay for the now excess supply. The question was also raised, who is going to build the power plants required by 2010 to meet growing demand?

Nationwide there seems to be a lot of anxiety over transmission facilities. Much of it surfaced in 2004 because of the blackout that occurred from Ohio to New York. The call was issued to reinforce the transmission systems' relay protection. The primary function of transmission systems is to transport energy. In transporting large amounts of energy great distances over transmission lines, potential problems must be anticipated because the system operates on the ragged edge. If adequate generation is to be provided within reasonable distances to loads, then less reliance is placed on the transmission systems. After all, generation is required before there can be utilization of transmission systems to get the energy to the loads.

Since the mid 1970s, our nation has had a critical need for a comprehensive energy policy. The U.S. continues to remain at the mercy of foreign oil, and now has become increasingly dependent on foreign natural gas and even foreign gasoline. One wonders what will be our national policy regarding the level of the nation's dependence on foreign energy?

With natural gas fast becoming the fuel of choice, many of the natural gas producers have taken to importing foreign natural gas in the form of liquid natural gas, or LNG. The nation's natural gas industry experienced a tumultuous period when the major petroleum companies backed away from domestic exploration and production of natural gas. Major companies, including ChevronTexaco, BP, Royal Dutch/Shell, and ExxonMobil, reduced their domestic drilling activity in favor of international exploration in countries such as Indonesia, Nigeria, Qatar, and Russia, thereby increasing our dependence on foreign energy. The recovered gas is converted to liquid natural gas for export to the U.S. However, one important impediment to importing LNG is the lack of adequate national facilities to receive and store it. The U.S. also imports gasoline because of limited refining capacity. Those imports, along with our domestic supplies, now meet our current energy requirements, but if a hiccup occurs with our foreign suppliers, we could experience a replay of the California 2000-2001 energy crisis—nationwide.

Vice President Dick Cheney predicted fossil fuels would remain America's primary energy source for "years down the road," and the current energy strategy should be to promote the increase of the supply and use of fossil fuels. In 2000, the DOE released 30 million barrels of crude oil from the nation's Strategic Petroleum Reserve to meet the nation's energy demand, but that raises the question, has it been replaced? Cheney argued the notion that we could simply "conserve our way out of an energy crisis" was not a practical solution to the nation's energy shortfall. He believes the government should promote new domestic sources of oil and natural gas. He was absolutely right about conservation—it is a mitigating contribution, not an answer to the energy crisis.

Additionally, he felt there should be a purposeful effort to build power plants. He projected a need to construct one new generating plant a week for twenty years. California's reliance on its neighbors for 20 percent of its power along with 30 percent of its own generation down for maintenance and/or repairs demonstrated how much the state needs to bring itself to be self-supporting in its energy requirements, avoiding reliance on its neighbors. Its own generation must be adequate to provide a reserve. California needs to evaluate the reliability of its existing generation, especially the older plants. Some of the older plants are over thirty years old and need to be retired. The state's electrical energy supply is a considerably larger problem than most people can fathom. California's 2002-2001 crisis stimulated the CEC to expedite the licensing of power plants. It adopted a "one-stop-shop" for permits, resulting in the licensing of nine big power plants totaling 6,278 Mw. If the CEC can license plants that quickly in a crisis, surely it should be able to shorten the licensing process during normal times.

The Bush administration also emphasized the need to open protected lands to oil and gas exploration, roll back environmental rules that constrain the burning of coal, and proceed with the construction of pipelines and refineries. Cheney emphasized drastic measures were required because the need is so great. He estimated there was a need for 38,000 miles of natural gas pipelines. Unfortunately, the energy bill laid dormant in Congress.

Vice President Cheney said building new nuclear power plants and extending of the life of existing nuclear power plants was the most environmentally satisfactory way to increase energy supplies. Environmentalists contested that claim; they argued that mining and the enriching of uranium along with the storage of the nuclear waste were environmental hazards. It appears the environmentalists will oppose almost anything that will contribute to the nation's energy supply and will offer no realistic alternatives. They will one day have to recognize that some hard choices will have to be made and if not now, then by future generations.

A Field Poll released May 22, 2001, indicated that fifty-nine percent of Californians polled favored building more nuclear power plants in the state. Opponents said the poll results might be misleading, as the nuclear issue had been dormant for so long. Seventy percent of those polled favored capping wholesale prices. The poll also showed Californians in favor of relaxing air quality standards to allow older power plants to be upgraded. Californians were not particularly enthused about the sale of $13.4 billion in bonds to pay for electricity and rejected the idea of using eminent domain to seize power plants. The majority of registered voters opposed additional offshore drilling for natural gas and oil to ease the energy crunch, but apparently, they would accept nuclear generation in lieu of drilling.

All these problems emphasize the need for comprehensive state and national energy policies. And the environmentalists and NIMBYs need to come to the party. Environmentalists have been dominating the scene for years. It is important to remember that California not only had an electrical energy production shortage, it had a natural gas shortage, and the state relied on its geographic neighbors to make up the shortfall. Couldn't we help solve the state's natural gas shortage by more exploratory work in California, possibly with offshore drilling along California's coast? Environmentalists have blocked drilling in ANWR, along the California Coast, and in the Gulf of Mexico. While the environmentalists have a point and probably do make a contribution to the quality of our lives, the general public has to understand that we, as a nation, eventually must be prepared to make some hard decisions in regard to the quality of our lives, in terms of seeking a balance between our energy requirements and the environmental impact on the quality of our lives. The longer the public puts off making those decisions, the more future generations will be faced with making those hard decisions and the greater will be the impact on their lives. A comprehensive energy policy must establish what the ground rules will be and what the trade offs will be between the need for new energy and the environmentalists' proclivity for protecting our environment.

Notable efforts have been made to explore the possibilities of renewable energy sources. In 2001, the Regional Council of Rural Counties proposed reopening thirteen existing biomass facilities that together were capable of producing 150 Mw. California at the time had twenty-nine cogeneration plants burning wood and tree wastes producing 540 Mw. Biomass facilities that burn waste materials, farm prunings, and forest brush have limits on the amount of energy they can be produce; the process also raises many environmental concerns. Senator Feinstein recently urged the Lake Tahoe Basin to develop a plan that would qualify the Basin for funds from the Healthy Forest Restoration Act to remove deadwood and underbrush; the law had authorized $760 million for that activity. Senator Feinstein's objective was to prevent devastating fires in the South Lake Tahoe Basin similar to that which Southern California had experienced in the previous year that killed twenty-three people and destroyed considerable property. This activity could make a significant contribution to the development of biomass alternative energy. There are those who argue the logging industry will make money from this activity, but so be it—it will increase forest fire safety, and it will produce jobs benefiting the state's economy. Commercial enterprises are entitled to make a reasonable profit.

A very practicable method to increase the use of solar energy would be the application of solar energy to residences and commercial buildings by developers. Solar shingles and other devices are now available, and their application to residences and commercial facilities by developers could have a significant impact on the demand for electrical energy. Additionally, solar energy can be used for hot water and space heating. Both of those applications are practical and achievable. To force this issue, solar energy equipment maybe should be required upon application for building permits, with the cost to be included in the price of the home. Yes, it will increase the cost of the home, but that cost would be made up over the life of the home, normally thirty years, with no, or minimal, ongoing energy utility bills.

Geothermal energy can make a significant contribution to California's energy needs and is a reasonably reliable source of energy. The impediments to its development are the drilling of wells and the construction of power plants. Developers are faced with the agony of convincing the environmentalists, NIMBYs and governmental agencies of the viability of the resource and convincing the folks it will have minimal impact on the environment.

Wind was considered a viable solution to the state's energy requirements, but it too has its impediments: location, unsightly structures, the killing of raptors, and the acreage required for wind farms. One has to consider the fact that wind energy production occurs only when the wind blows. In the 1970s, the Friends of the Earth contested the development of wind farms on Altamont Pass. How about wind turbines along the coastlines as they do in Europe? Germany is now building 5 Mw wind turbines for offshore installation. The structures have two hundred-foot blades and stand six hundred feet tall.

All this being said, however, it is should be readily evident that while renewable energy would make a significant contribution to the state's energy requirements, but it is not a total solution. Except for biomass and geothermal facilities, renewable facilities are not reliable sources of energy throughout the day and require reliable backup generation to assure firm deliverable energy.

The idea that we can conserve our way out of the energy crisis is also an unrealistic illusion. Conservation can and will make a large contribution to our energy needs, if properly handled, but it is not a panacea; we will still need a substantial amount of generated electrical energy. There will always be a need for conventional power plants to assure a continuous and adequate supply of firm electrical energy to meet consumers' immediate demand.

The CEC predicted that by 2004 California would have a power surplus. CEC's Steve Larsen expected all of the twenty-one power plants that had been licensed would be built, but industry executives believed only about half of those plants would actually materialize. That revives the question, how do we make reserve generation energy available in a free market? One possible scenario would be using long-term contracts for reserve capacity that could be called upon in emergencies, but the viability of that idea is questionable.

Jay Hakes, former director of EIA, observed that the problem was not that energy companies could not respond to demand, but that the newly deregulated energy companies were now denied the certainty of earning a return on investment, which state and federal regulators once provided when the utilities were regulated entities. He properly identified the problem to be with the deregulated energy companies—lack of certain return on investment. He believed Washington was bereft of solutions to the real energy problem, saying that deregulation cut the fat off the system and left people vulnerable to very volatile markets.

What are the incentives for generators to keep ahead of the growing demand for electrical energy? Generally they barely keep up with demand without providing any reserve. Prior to deregulation, the regulated utilities, in their obligation to serve, recognized their responsibility to plan ahead for supply reserves to their respective service areas, and the cost of those reserves was built into the rate. Deregulation relieved them of the responsibility to provide the necessary reserve that would assure reliable service—that responsibility was transferred to the ISO and the energy market. For reserve energy then, the ISO had to rely on the energy marketers, suppliers who had no incentive to cooperate.

In drafting AB 1890, considerable rhetoric had focused on the concern that the state's three major utilities would find ways to keep wielding monopoly power over the consumers. Thus, the restructuring set out to forestall that possibility by forcing the utilities to sell their fossil fuel power plants and allow new generators to enter California's energy market. The utilities then had to rely on the wholesale market for energy to adequately meet the consumers' requirements.

FERC eventually realized that California's unrestrained energy market really needed attention. FERC's chairman dedicated nearly one hundred people, a tenth of FERC's budget, to a new office charged with addressing the daily activity of the power markets nationwide. This new Office of Market Oversight and Investigations had responsibilities that included identifying market manipulations and tracking daily energy markets nationwide, hopefully to provide a safety net to the states with deregulated markets, including California. The Bush administration and Congress wanted FERC to get off the dime and address the allegations. The question is, in what way were the new Office of Market Oversight and Investigations responsibilities different from FERC's original mandate?

Senator Burton introduced a bill to create a California public power authority, including the authority to sell revenue bonds to finance the construction of new power plants—a knee-jerk reaction to the crisis. The energy crisis was immediate, and establishing a power authority and constructing power plants would take years to implement and produce power, years in which the private sector could have easily built the necessary power plants without the added bureaucracy. Burton also suggested the

State should acquire some of the utilities' assets with the comment, "I give you a dollar, and you give me a hot dog." Burton apparently didn't realize he already had his "hot dog" in the form of high-priced electrical energy.

Phil Angelides had a similar knee-jerk reaction. He proposed a plan where the California Consumer Power and Conservation Authority (now the California Power Authority) would be allowed to issue $10 billion in State bonds that would finance conservation programs through low interest loans, finance the purchase of the utilities' transmission systems, and authorize the building of power plants. A conservation program would have provided some immediate relief, but the acquisition of transmission facilities would not have provided any additional energy that was so badly and immediately needed and, again, with the years required to build power plants and take over transmission, the plan would contribute nothing to the crisis at hand.

Apparently, there was a "securitization" plan floated to which the legislature gave little support. The plan was for the State to sell bonds and retire the utilities' debt. The ratepayers would repay the debt. When you think about it, that approach to retiring the debt would have been a sensible solution; it would have paid for high-priced energy already used by the consumer. PG&E would not have gone into Chapter 11 bankruptcy. In hindsight, had the retail rates been allowed to float along with the wholesale prices, the utilities' debt would not have occurred or would have been substantially reduced. Energy would have been paid for as it was consumed, and the consumers would have carried their share of the burden. The consumer advocate groups should have recognized that. As it is, the ratepayer is now, and will be well into the future, paying for that expensive energy at long-term contract energy prices.

Obviously, the crisis occurred because there was no one "at the helm." What we had was a number of independent departments and organizations each exerting some form of control over different elements of our energy usage. California needs an energy department headed by an energy czar with a complete chain of command that incorporates all of the existing energy departments and organizations, and that would include the PUC, the CEC, the ISO, the ARB, and the DWR. The czar should report to the governor.

Under the suggested organization there would be accountability. As it is, there is no accountability for California's 2000-2001 energy crisis. The PUC, along with FERC, should accept accountability for pushing the utilities into enormous debt. However, they are not responsible for the crisis as the crisis was caused by a shortage of energy, a shortage they had no control over. The CEC should have warned of the danger if the state lost its ability to import power. But then maybe they did. They had no authority to construct power plants; their authority was limited to licensing them. The ISO was limited to scheduling power, which it did a fairly noble job of. The ARB's role was that of controlling emissions; they had absolutely nothing to do the amount of energy used or supplied. The DWR was called upon to purchase energy for the beleaguered utilities and maintain them as viable entities. So it appears the only one that should have accountability was the governor, and he lacked the primacy to take control, but is he really at fault? He lacked an energy czar, a person to inform and advise him, who could have laid out the full picture for him, which could have resulted in immediate decisions to avoid the pending crisis. Let me repeat—energy is the essence of our economies; we need to drive that thought home.

Deregulation for California, we can say, "has been tried," and "fool me once . . ." The State has to keep the people's interest in the forefront. It has been demonstrated that the energy suppliers agenda is not looking after the people's interest—their agenda is their own profit, and there's nothing wrong with that. However, it doesn't belong in a commodity as crucial to the people's interest as electrical energy. California's crisis adequately demonstrated that.

California should return to being a utility-regulated state. Utilities should be required to service their customers with reliable electrical energy, a process that includes generation, transmission, and distribution of energy, with sufficient reserves to maintain firm power. And the state should become totally self-sufficient with regard to electrical energy.

The utilities that provide power generation must guarantee adequate generation to service their customers. The power generators can participate by contracting with the utilities to sell their power. If a utility decides an additional new power plant is required, the utility should determine the size and location in consultation with the ISO and the CEC. The utility will then select the site and apply to the CEC for certification and proceed through the certification process. Upon certification, the utility, under the auspices of the PUC, puts the plant out for bid in which the utility will submit its own bid for construction of the plant. The bids are to include the spent cost of site selection and plant certification, which will be included in the customer's energy rates for that plant. The PUC will evaluate the bids and select the winning bid. The criteria for the plant site selection should be its physical relationship to the loads, available transmission facilities to accommodate the plant and, if necessary, the availability of cooling water.

The PUC will monitor the utilities' costs and in establishing its rate structure, the PUC must adhere strongly to the requirement that rates will allow the utilities to recapture their costs plus a reasonable return on investment. The PUC's penchant for keeping electric rates low to the determent of the utilities must not happen again.

One final comment before closing this chapter: there is much to-do worldwide regarding global warming, and it is said to be caused by "greenhouse" gases that are released upon the combustion of fossil fuels—oil, natural gas, and coal. Environmentalists have demanded that power plants be required to mitigate the release of combustion byproducts for the benefit of human breathing, ecology, and to prevent global warming. I concur with those objectives, and I agree that there is global warming—I just don't believe that greenhouse gases cause it. Carbon dioxide (CO_2) is held to be the primary culprit and, therefore, efforts are being made to have power plants capture and sequester CO_2 to reduce global warming, an effort I believe to be an exercise in futility.

IX. Daybreak— Meeting California's Energy Challenge

As I have continually emphasized, *energy is the essence of our economies.* Our nation's and our state's economies have become critically dependent upon the availability of electrical energy. And not just our economies in the narrow sense, but our lifestyles and culture have come to rely completely on electrical energy—a commodity that is absolutely indispensable and has to be available upon demand. When we walk into a darkened room and flip on a switch to have light, we are placing a demand on the electrical systems, and the fact that we're making a demand rarely enters our minds.

We all enjoy the fruits of a robust economy. Electrical energy being the essence of our economies, it is therefore imperative that we maintain our electrical systems as reliable as possible; however, we must recognize there is a cost for that reliability. Those costs are just not financial; they also include impacts on our environment and on our culture. Much can be done to mitigate those financial and environmental impacts and to comfort the NIMBYs; however, it has be emphasized that electrical energy comes with a cost.

Up until 1998, California's energy market had been regulated under the supervision of the California PUC, which had worked fairly well. In 1998, Assembly Bill 1890—utility deregulation—was activated. It required the transformation of California's electrical services industry from a regulated market to a quasi-competitive market. The objectives of the deregulation were *to lower the cost of electricity,* to retain and attract jobs, and to reduce power outages. Until the years 2000 and 2001, 20 percent of California's electrical energy was imported from neighboring states. Californians assumed that this arrangement would continue indefinitely, and this assumption was a major factor in California's failure to build its own power plants. In 2000-2001, the neighboring states unexpectedly needed their formerly surplus power, and California's electrical industry suffered a chaotic energy crisis. Our regulatory agencies' exacerbated the crisis by failing to meet their fiduciary duty to properly control wholesale and retail electrical rates. The result was the cost of electricity went out of sight. The utilities went into enormous debt, jobs were lost as large industrial and corporate energy users moved out of the state, and California endured crippling power outages—completely contrary to the original objectives AB 1890.

In a competitive market, the cost of commodities and services will naturally increase when supply fails to meet demand—a financial nuisance to be endured until supply catches up with demand. But, because of the economy's vital dependence on electrical energy, a sustained shortage of electrical energy cannot be tolerated. Adequate supply along with a reserve of electrical energy must be constantly available, and a competitive market will not guarantee that. The power generators will produce only enough power to meet existing demand, if the price is right. The utilities that operate in a regulated market have, by law, an obligation to serve. In meeting that obligation in a regulated market, the utilities planned ahead to provide adequate supply of energy along with a reserve. AB 1890 relieved them of that requirement. It was simply assumed that utilities in a competitive market would automatically live up to their public responsibility, but the competitive market provided no incentive for them to do so.

It became apparent that a truly competitive market would not provide a totally reliable continuous source of electrical energy and should not be expected to do so. Therefore, the electrical energy market must be, at least, a quasi-regulated market; a degree of regulation will always be required to properly meet the public's electrical energy requirements.

Rates in a regulated market may be higher than those in an unregulated market, but what the consumers are buying with that higher price is the assurance of sufficient energy to meet their immediate demand from a utility that plans ahead to meet that demand. I believe that extra cost is well worth the comfort it provides.

The retail electrical commodity market will always have to be regulated, except for those consumers who elect direct access. However, those who elect direct access must be faced with certain restrictions. Once they elect direct access, they should be required to remain in the direct access mode and not allowed to return to the utilities for a given period of time, say one year or more. If they do eventually return to the utilities, they again should be required to remain with the utilities for a given period of time, again maybe one year or more. These restrictions would prevent the large consumer from jumping in and out of utility service, action that would impede the utility's ability to plan and provide reliable service.

A degree of competition can be allowed in the wholesale market. The utilities should be allowed to choose to build power plants and/or contract for power. Given this choice, the utilities would again assume their obligation to provide adequate and reliable service. At the same time, the independent power generators would participate in power production and would compete with the utilities. The utilities should remain regulated by the PUC, whose oversight should include the utilities' decisions to build power plants and/or contract for power. The option to contract for power should not be unreasonably withheld as it has been in the past. This would essentially eliminate FERC's authority over the production of power; however, the FERC would still retain its authority over the production of hydroelectric power and the transmission of power.

The state's PUC's should continue to oversee service standards, safety rules, and designs. It should continue to authorize utility rate changes once control over power generation and/or the purchase of electrical energy is returned to it. It should continue to oversee markets to constrain any anti-competitive activity. The PUC's prosecutorial authority should remain for unlawful utility retail marketing and billing activities, including business relationships between utilities and their affiliates.

The Federal Energy Regulatory Commission should return to being an independent federal agency responsible for regulating the interstate transmission of electricity, natural gas, and oil, and for the management of hydroelectric projects. With the exception of control over hydroelectric power, it would lose its authority over the production of power.

With the advent of California's deregulation under AB 1890, the State was required to establish an Independent System Operator overseen by FERC. As a result, FERC essentially gained authority over the production of power for the state. Given FERC's authority over the California ISO, it also gained authority over investor-owned transmission. Apparently, FERC has that same authority over other parts of the nation. The FERC now wants more power over the western states' electrical systems—it wants to incorporate the California ISO into a regional ISO. FERC's dismal performance during California's energy crisis of 2000-2001 makes it a very poor candidate for controlling the state's ISO and the production of electrical energy. California would be better served by retaining control over its own electrical systems and becoming self-reliant in the production of power. The ISO serves the state well in the management of the state's transmission systems and the scheduling of generation to meet load requirements.

FERC's plan to standardize electricity trading across the contiguous continental United States would give FERC increasing control over pricing, planning, and operating decisions of electrical systems. It's obviously a power grab on the part of that administration. Under its new role, FERC would have to coordinate expanded and complicated transmission systems, which would probably require complicated rate structuring. Transmission systems already have had their problems—consider the complete shutdown that occurred in 2003 across Canada, Ohio, Pennsylvania, and into New York. Additionally, one wonders would FERC's plan make the transmission systems more vulnerable to terrorist activity? Generation close to load centers makes a lot more sense than expanded transmission systems with remote generation; distributed generation would be a step in the right direction.

The population of California is constantly increasing and each new person places an additional burden on the need for food, housing, jobs, transportation, and electrical energy. The following table with data from United Nations and Bureau of Census gives the world's human population growth in millions, in fifty-year intervals, since the year 1750.

Year	Population Growth
1750	790 million
1800	980 million
1850	1,260 million
1900	1,650 million
1950	2,555 million
2000	6,080 million

The earth's population in 2004 was reported to be 6.4 billion. It is obvious the earth's population is growing exponentially. In the year 2050, following the current trend, can we expect the earth's population to double to about twelve billion people increasing to about eighteen billion by the year 2100? California can be expected to at least follow the same population trend, and that population growth will place a phenomenal demand on the state's electrical energy systems.

Having said that, and after witnessing an energy crisis in California (the fifth-largest world economy), along with witnessing a multi-state blackout in the eastern part of the nation, it is obviously time, and an absolute necessity, that our governments, federal and state, establish long-range, comprehensive energy policies and that they do it without delay. They should develop policies that will provide for the state's and the nation's energy requirements well into future for generations to come. The policy must provide a firm assessment of what energy resources are available, or can be made available, in what quantities, and how best to utilize them. The policies should include assessments of any associated environmental impacts with recommendations for acceptable tradeoffs. Once having established those policies, California would then be equipped to move forward in an organized manner to develop reliable energy systems to meet the state's ever-growing, voracious energy appetite.

It is my recommendation that California become completely self-reliant and that California establishes its own electrical system isolated from the rest of the nation, including having its own ISO not regulated by FERC. In that way, California can have complete and total control over its electrical requirements. In fact, I would recommend all states adopt that policy.

The most effective way of producing effective and reliable electrical systems requires a complete understanding of the total system, its operation, its requirements, and its environment. That understanding includes knowing current loads and demands, future load projections, available energy resources, status of the existing system's maintenance, and the improvements necessary to maintain an effective and reliable electrical power system. The oversight of energy systems should be under the watchful eye of an energy czar.

The California ISO, in conjunction with the CEC, manages the state's electrical transmission systems and schedules power generation to accommodate loads, so it probably knows the current status of the state's electrical systems. What it doesn't know for certain is how that load will vary from time to time; much depends upon the weather, business operations, business and industrial development, and residential development. Also, what the ISO does not know for certain is how the state's loads will grow, by how much, where and when to expect those loads, and where and what will be the generation required to accommodate those new loads.

The current condition of the existing systems may be in question and is probably a quagmire of unknowns, as there are a multitude of owners. The utilities, as directed by AB 1890 deregulation, were required to sell many of their fossil fuel power plants to several buyers. Many of those plants were, and are, thirty years old or older and have probably outlived their useful productive life. One wonders, has the State supervised the maintenance of its electrical systems? Does the CEC have that data? What is the status of upgrades for those plants, and how far into the future can they be considered reliable? What is the projection of load growth in the future—how much, where, and when? Perhaps the CEC has that information.

The State currently has a pretty fair understanding of the quantity of energy available through the long-term contracts that the DWR negotiated at relatively high prices, contracts that kept California supplied with energy during the crisis and that the ratepayer continues to pay for. But, what will be the resources beyond those long-term contracts? Is there adequate generation to supply California with firm power, including reserves? The state has in the past depended on its neighbors for 20 percent to 30 percent of its requirements, and that has not included any allowance for reserve capacity. The current call is for a 15 percent to 17 percent reserve capacity with no viable plan on how to achieve and maintain that reserve. Currently, the PUC requires the utilities to project and contract for 90 percent of their electrical energy one year in advance, and supposedly that includes a reserve. Apparently, many are expecting the competitive market to supply the reserve. That is wishful thinking. As the current market will meet only what the market demands, there has to be an incentive for generators to supply reserve capacity. Presumably, in their one-year contracts, the utilities will offer some form of compensation for standby reserve, which probably will result in take-or-pay contracts.

The transmission systems within the state are said to need upgrades. Those systems are also old, and there was the Path 15 incident during the crisis in which congestion problems were experienced. Again, there are multiple owners of the transmission systems, and those systems are managed and coordinated by the ISO, with each owner responsible for the maintenance and upkeep of its own systems.

Distribution systems are the responsibilities of the utilities, so the utilities generally know the status of those systems and their state of repair.

Currently, the state's energy planning appears to be flying by the seat of its pants. The author recommends that the State of California perform a thorough and detailed statewide power system study that would (1) assess the status of the state's current electrical systems, (2) determine how well those systems are equipped to meet future needs, and (3) develop a strategy to meet those needs. A short-term time frame of thirty to forty years should be the immediate consideration, with longer-term system estimates of, say, one hundred to one hundred fifty years. It is recommended the study be performed as a special project by the CEC in cooperation with the ISO. The author recommends the utilization of the private sector for the detailed information gathering study work. The study should be under the oversight of a governor-appointed energy study committee to be made up of one representative from each of the following, along with a governor-appointed chairman:

- The three private utilities
- The three major public utilities
- California Energy Commission
- Independent Service Operator
- Air Resources Board
- Sierra Club
- A construction expert
- A rate expert
- A financial expert
- A demographics expert
- Legal counsel
- FERC
- Public Utility Commission

Upon completion of the study, the committee chairman could be appointed as the state's energy czar with the authority to oversee the state's total energy systems and resources.

The first responsibility of the committee would be to search for and employ a project manager. That person could either come from the CEC staff or be a completely new hire. The objective would be to obtain the most talented and experienced individual for this major assignment, one who is business wise and technically oriented in the energy business. The project manger's first task would be to establish and organize a task force to perform a suggested functional organization is identified in Figure 6, California Study Organization Chart, below.

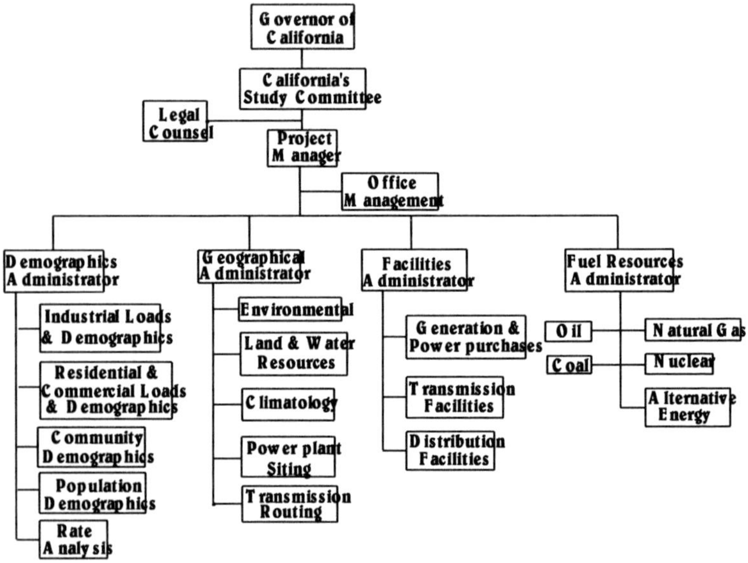

Figure 6.
California Study Organization Chart

Additionally, the CEC could be asked to provide an office manager whose initial responsibility would be the logistics of office space, furniture, equipment, supplies, and office organization.

All fieldwork activity requiring data gathering would be performed by private sector consultants who are supervised and managed by a study staff, many of whom might be drafted from the CEC. The CEC study staff members would manage, review, and correlate the data for their particular assigned tasks. They would also make recommendations based on their evaluation of the data and would prepare a final comprehensive state energy report. The study will include the following activities as appropriate:

Industrial, commercial, and residential loads and demographics: A review of all major industrial, commercial, and residential loads, including load size, projected expansions and/or reductions, location, and density

Community demographics: A study of cities and townships and their plans and attitudes toward industrial and commercial expansion and development within their purview

Environmental: Interaction with various environmental groups to determine their concerns and attitudes towards the production, transmission, and distribution of electrical energy, including the siting of facilities

Rate analysis: A review of the history of rate structuring with projections of future rates under the new organization, including interaction with consumer advocate groups

Land and water resources: A search and analysis of potential power plant sites and transmission routes and what water resources are potentially available for cooling water

Meteorology: Weather trends, rainfall, and snow pack history, including possible projections of hydroelectric generation

Projected generation requirements: Predictions of future generation required to meet load growth

Recommended generation methods: Recommendations for generation methods with emphasis on alternative and distributed generation

Power plant siting: Identification of optimum generation sites to accommodate loads

Transmission routing: Search for optimum transmission routes, including geography and consideration of right-of ways

Generation and power purchase contracts: Identification of existing generation, size, type, location, age, state of repair (including review of maintenance records), and expected life, as well as expiration of existing long-term purchase contracts and their replacement

Transmission facilities: Review of existing transmission systems for capacity, routing, state of repair (review of maintenance records), and proposed improvements

Distribution systems: Review of existing distribution systems for capacity, routing, state of repair (including review of maintenance records), and proposed improvements

Fuel resources: An overall review of available fuel resources, fuel transmission facilities, storage, current utilization, future projections, and any waste disposal (primarily nuclear)

Associated load study projection calculations: Additional data as necessary.

Based on the report's recommendations, the committee would then develop the state's overall energy policy.

I have estimated the study would require a commitment of $42 million, scheduled over a three-year period (see Appendix C). I believe this estimate is adequate to carry out the study. The estimate was organized as follows: The study committee would be compensated for their attendance at the monthly meetings at the rate of $220 per hour for eight hours at each monthly meeting. Legal counsel was allocated ten hours per month at an hourly rate of $440. The project manager was allocated $11,000 a month ($132,000 a year), as this is a critical position. An office staff of six people was estimated to support office operations each at an average rate of $3,300 a month. The administrators were allocated $7,370 a month ($88,440 a year), and the contract mangers were estimated at $5,500 a month ($66,000 a year). Facilities expense is self-explanatory. The study contracts were each estimated at $88,000 a month for 18 months. A 10 percent contingency for the total study was included.

Following the above study proposal, California would have, in thirty-six months, a clear analysis of the state's standing in relation to its energy requirements and would be well on its way to designing an energy policy to meet future energy requirements to bolster its dynamic economy.

In conclusion: California has the opportunity to demonstrate progressive leadership to the nation and to the world—if the state has the courage and commitment to take the bull by the horns and act. It will require strong leadership—will Governor Schwarzenegger provide it?

Having said all of this, I conclude this book. *Is anybody listening*?

Appendix A. California Utilities

2002 California Energy & Power
Courtesy of the CEC

IN-STATE PRODUCTION	GWH	%	MW
Coal	27,618	10%	3,153
Large Hydro	26,937	10%	3,075
Natural Gas	90,898	33%	10,376
Nuclear	34,353	13%	3,922
Renewables	29,879	11%	3,411
Biomass	7,140	3%	815
Geothermal	13,946	5%	1,592
Small Hydro	4,382	2%	500
Solar	864	0%	99
Wind	3,546	1%	405
Total In-State	**209,685**	**77%**	**23,937**
NW IMPORTS	**GWH**	**%**	**MW**
Coal	5,283	2%	603
Large Hydro	19,304	7%	2,204
Natural Gas	1,717	1%	196
Nuclear	882	0%	101
Total NW Imports	**27,186**	**10%**	**3,103**
SW IMPORTS	**GWH**	**%**	**MW**
Coal	21,582	8%	2,464
Large Hydro	1,959	1%	224
Natural Gas	6,865	3%	784
Nuclear	5,267	2%	601
Total SW Imports	**35,673**	**13%**	**4,072**

GRAND TOTAL	**272,544**	**100%**	**31,112**

Appendix B. PG&E's Power Content Years 2002 & 2004

Power Content Label

ENERGY RESOURCES	**PG&E 2004 POWER MIX** (Projected) 0% of power purchased by PG&E specifically purchased from individual suppliers	**2002 CA POWER MIX** (For Comparison—Projected) Percentages estimated annually by CEC based on previous year's consumption
Eligible Renewable	13%	9%
Biomass and waste	4%	3%
Geothermal	3%	4%
Small hydroelectric	4%	2%
Solar	0%	0%
Wind	2%	<1%
Coal	3%	15%
Large Hydroelectric	19%	23%
Natural Gas	42%	42%
Nuclear	23%	11%
Other	0%	0%
TOTAL	100%	100%

Data taken from enclosure in PG&E monthly energy bill

Appendix C. California Power System Study Estimate

10/09/2007

PROJECT DURATION: 3 years

SCHEDULE

Project Organization: 6 Months

Duration of Contracts: 18 Months

Analysis & Report Preparation: 12 Months

Expense Category	Function	Qty	Rate	Hours	Months	Cost
STAFF	Study Commit-tee	18	$220/hr	8	36	$1,140,480
	Legal Counsel	1	$440/hr	10	36	$158,400
	Project Manag-er	1	$11,000/mo	FT	36	$396,000
	Office Staff	6	$3,300/mo	FT	36	$712,800
	Administrators	4	$7,370/mo	FT	36	$1,061,280

	Contract Managers	18	$5,500/mo	FT	36	$3,564,000
Staff Subtotal						$7,032,960
+30% PR Burden						$2,109,888
Subtotal						**$9,142,848**
FACILITIES/AUTO	Furniture	23	$2,200/office	-	-	$50,600
	Equipment: Phones, Copiers, FAX, Tools, Supplies, etc.	As needed	-	-	-	$25,000
	Laptop Computers	23	$2,200/ea	-	-	$50,600
	Mainframe Computer	1	$4,400/ea	-	-	$4,400
	Automobiles	23	$33,000/ea	-	-	$759,000
	Office Rent	5,000 sq ft	$1.54/ sq ft/mo	-	36	$277,200
	Travel: Hotel	23	$726/mo	-	36	$601,128

	and Food					
	Auto Expense: Gas, Oil, Lube	23	$189.20/mo	-	36	$156,658
Subtotal						**$1,924,586**
CONTRACTS	Consulting Firms	17	$88,000/mo	-	18	$26,928,000
Subtotal						**$26,928,000**
Project Total						**$37,995,434**
+10% Contingency						**$3,799,543**
Grand Total						**$41,794,977**
TARGET						**$42,000,000**

About the Author

Bill Dulude is a retired professional electrical engineer who has spent over forty years in the electrical engineering design and construction industry. His forte has been experience in heavy power system studies, including report writing, estimating, design management, construction engineering, and construction management of major projects. He has a predilection for electrical energy.

Bill was born February 14, 1927, in a small town in Minnesota where his father was a physician and surgeon and Bill was one of six siblings. He was orphaned in 1940, thereafter spending three years in Oregon with his half-sister and then one year in Minnesota living with his older brother and his wife. In 1944 he quit high school after his junior year and joined the Navy so as not to be a burden on his brother's family. In 1947 Bill entered the California State Polytechnic College (now University) in San Luis Obispo, California majoring in electrical engineering. Upon graduation he joined Bechtel Corporation, and in Bechtel's employ Bill performed construction engineering on a number of major projects, including the original 1,200 Mw Pittsburg Steam Plant in Pittsburg, California; Aden Refinery in the country of Aden; Union Oil Decoking Plant in Guadalupe, California; Shell Oil Refinery in Anacortes, Washington; Mobil Refinery in Beaumont, Texas; Zirconium and Titanium Ore Reduction Plants in Ashtabula, Ohio; Fiberglas Yarn Plant in Akin, South Carolina; and General Foods Plant in Dover, Delaware.

Upon completion of the Aden project he married Barbara Hewitt, who has been his bride of fifty-three years. They raised two sons, Mike and Doug. In the '60s he and his family spent eighteen months in the Philippines in the employ of a consulting engineering firm. While with that firm, Bill participated in a combined power system study for Subic Bay Naval Base and Cubi Point Naval Air Station in the Philippines. Upon his return to the states, he managed very successful power system studies on two major military bases, one for the Alameda Naval Air Station in California and one for the Shemya, Alaska Air Force Base.

Subsequently Bill joined another engineering consulting firm as vice president, eventually becoming the firm's senior vice president. During that employ he supervised the engineering, project management, procurement, and construction management of four geothermal power generating projects located in Managua, Nicaragua, in Tibet and two in the geysers area of California. The projects ranged in size from 3 to 110 Mw.

Bill retired in August of 1989. In his retirement, he witnessed the California 2000-2001 energy crisis. Bill followed the crisis with avid interest and decided to write a book about it. *Shades of Darkness: California's Energy Crisis* is the result of his labors.